The Diary of a Claines Farmer
1883

Isaac John Sansome

of Oak Farm, Bevere, Claines, Worcestershire

Transcribed from the original and edited by

Geoff Sansome

Foreword by

Nick Evans, Professor of Rural Geography;
Director, Centre for Rural Research,
University of Worcester

and

Hugh Mercer Curtler, Emeritus Professor of Philosophy
Southwest Minnesota State University

With original artwork by John Coleman, Claines

The Diary of a Claines Farmer 1883

First edition printed 2017 in the United Kingdom

Published by Geoff Sansome

For more copies of this book, please email: Sansomes@Hawford.Farm

Tel: 01905 754561

Printed by Sprint, Cambrian Printers

ISBN 978-1-78808-205-1

Through buying this book you are supporting the work of the Farming Community Network.

CONTENTS

FOREWORD

By Nick Evans and Hugh Mercer Curtler

Nick Evans

Against a backdrop of a multitude of social, economic, environmental and political pressures, the continued survival of the family farms in Britain today remains a remarkable phenomenon. If economic theory is to be believed, they should have disappeared many years ago (just as have most small motor car manufacturers)! This raises the central question as to why such forms of production have proved to be so resilient over time. Geoff Sansome's *Diary of a Claines Farmer* goes some considerable way towards providing an answer. The determination to 'keep the name on the land' and follow in ancestral footsteps, as the book itself proves, shows that social and cultural factors have operated in extremely powerful ways over the centuries. The enormous lengths to which farmer Isaac John Sansome goes as he strives to make some kind of living, throughout every week of every month illustrated in this diary, demonstrates that profit alone cannot account for the existence and structure of farm family life.

Indeed, whilst historians often out of necessity are selective of events in their portrayals of the past, the *Diary* provides a complete late 19th century picture of farm life in central Worcestershire. It proves that the ordinary aspects of life, which are so frequently omitted, are as fascinating as the extraordinary. The geographical range of Isaac John's contacts may be limited mainly to within the county, but their number is remarkable, as is the mobility shown in an age before motor vehicles. Therefore, the *Diary* serves to dispel any pre-held notion of an insular, sedentary and simple farming life, as we look back through rose-tinted spectacles and sanitise the past. Perhaps the most striking aspect of the *Diary* is its emphasis on the 'informal economy'; one based upon social and cultural capital, well-oiled by the availability of beer and cider! By providing direct transcript, the subtleties of day-to-day farming existence are captured in full.

The emergence of an 'agricultural census' from 1866 begins to provide us with some statistical data about the farming industry in Worcestershire in later Victorian times. However, the approach taken in this book is to provide rich, qualitative 'data' to paint a picture that no numerical data can provide. It represents a detailed case study, one which has relevance far beyond the county boundary; hence it should appeal to everyone, from the casual reader to scholars of social history.

There are some familiar messages about the profitability of the farming industry and its demands on those involved, which serve to chime with the present, but the text can be read in numerous ways. Each reader will therefore take away something different from the *Diary*; that said, one message is clear: almost a century and a half on, the tenacity of family farms in today's world owes much to the efforts of individuals such as Isaac John Sansome. Long may they continue!

Nick Evans is Professor of Rural Geography and Director of the Centre for Rural Research, University of Worcester, England.

***

Hugh Mercer Curtler

The Diary of a Claines Farmer opens a door into the life of a hardworking and enterprising farmer in Claines, Worcestershire in the latter part of the 19th century. The period photographs included by Geoff Sansome, and his remarkably thorough and detailed footnotes, enliven the diary considerably and provide a helpful context. The daily notations of Geoff's ancestor, the farmer Isaac John Sansome, reveal to us the busy life of a man recovering from the premature death of his first wife, who left him with two small children. The diary provides a cross-section of his life, including subtle hints of a budding romance with his second wife.

As a descendent of Thomas Gale Curtler, the squire whose farm Isaac John rented, I found this diary quite fascinating. I was especially interested in seeing the many references to the squire's eldest son, the Rev. Thomas Gale Curtler, who seemed to be in constant need of renting Isaac John's "nag mare." In addition, I applauded Isaac John's determination to approach the squire in order to convince him to lower the rent on his farm — given the fact that in the year of 1883 the farm operated at a loss in excess of £75!

In all, Geoff Sansome has done us all a favor in shedding a light on a small part of the world. It was surely a simpler time, a time when folks worked hard, paid their bills, and interacted with, and relied upon, one another in ways we can only read about and envy.

Hugh Mercer Curtler Jr is Emeritus Professor of Philosophy at Southwest Minnesota State University and lives in Cottonwood, Minnesota, U.S.A.

PREFACE

This is the diary of my great-grandfather, Isaac John Sansome of Oak Farm, Claines, Worcester for the year 1883.

Whether or not this was the only year in which Isaac John Sansome kept a diary is not known. This single diary has been handed down within the family but only researched since 2015.

Isaac John was born in 1841 in Claines, Worcester, the son of an agricultural labourer. His first wife Sarah Ann died on 22 December 1882, aged 32. At that point they had two children, Rosamond Elizabeth (Rose), aged 11, and Jessie Hannah aged 2 ½. They had moved to Oak Farm, Claines in 1876. It was part of the 720 acre Bevere Estate and the owner landlord was Thomas Gale Curtler Esq, of Bevere House. At the time of writing Isaac was 41 and described himself as "A Farmer of 70 acres, pasture".

The diary is a factual account of the daily life of a small tenant farmer on the outskirts of the City of Worcester. Isaac John produced milk, traded in hay and fruit, hired out horses, bought and sold cattle and was a volunteer with the Worcestershire Yeomanry Cavalry. Isaac had a wide range of farmer and tradesmen contacts and the diary gives an interesting insight into the trades of Worcester at that time. It also illustrates the structure and prevalence of farming in Claines and the landlord tenant relationships in force.

Isaac John employed both permanent and casual labour when agricultural labourer was the main occupation for many and their reliance on beer and cider is clear!

As well as farming, Isaac rode with the Worcestershire Hunt; rode out to Worcestershire villages with friends and family for pleasure and enjoyed some of the local hostelries, with the occasional trip to the theatre. He appeared to be doing his best for his two children, who were being schooled and cared for by relatives in Staffordshire, as well as by his wife's first cousin, who lived in regularly at Oak Farm. Sundays were clearly for Church and family.

The diary records Isaac's main income and expenditure. It portrays a time of trust, barter, goodwill, borrowing and lending, with payments being made as and when they could be afforded. Isaac was regularly lending money to family and his acquaintances, as well as sending money to relatives in Staffordshire. The diary also subtletey and simply hints at a developing romance.

The diary itself is a small leather-bound pocketbook. Dates have been entered in Isaac's own hand and all of the entries are made in black ink, quite legibly save for the odd entry. Transcription was straightforward. Original spellings have been retained but punctuation has been added.

The date of the diary, 1883, has made it possible to reference the 1881 England census to research some of the names of the farmers, tradesmen and individuals mentioned. Local knowledge of Claines, its families and the outlying villages has helped piece together some of Isaac's contacts and movements. This additional information and any clarification of diary entries is provided as footnotes as the diary progresses. Only where it has been possible to research a name or place with any certainty have notes been provided.

I have provided a brief synopsis at the beginning of each month to help orientate the reader and a plan and map of Oak Farm as it was in 1883. Little has changed since then. For those not familiar with Worcestershire I have included a sketch map of the town, village and place names mentioned by Isaac, showing the extent of his regular travels within the locality.

A full list of names mentioned in the diary and brief details of these are given in Annex 1.

Based on the finances recorded in the diary a set of "accounts" for the year have been calculated and are presented. This does not capture all transactions as it is clear from the diary that all were not recorded. But it gives an interesting insight into Isaac's main sources of income and his expenditure. For readers not familiar with pre-decimal currency £5-10-6 is five pounds, ten shillings and sixpence, with 15/6 ½ being fifteen shillings and sixpence halfpenny. Footnotes explain the imperial system of weights and measures.

My thanks go to Hugh Curtler and Nick Evans for their contributions to the Foreword. Hugh and I met in 2007 when he travelled to England and I was able to help him discover his extensive Curtler family connections with Claines and its history. He is the 2nd great grandson of Isaac's landlord Thomas Gale Curtler Snr. Nick, in addition to his academic role, is Chair of the Worcestershire Chaplaincy for Agriculture and Rural Life which provides valuable connections, support and leadership in countryside matters across the county.

Transcribing the diary and researching the historical links kept me very occupied for many evenings and I would like to thank my wife, Ann, for her patience and support during my distracted times. Thanks also go to Rachel Cramp, the current churchwarden at Claines, for her painstaking proof reading and suggested edits to the diary. Thanks must go to my late Uncle, Peter Sansome, of Oak Farm for realising the significance of the diary and for passing it down through the family. Finally the wonderful water colour paintings of the farm animals are from an original painting by a dear family friend, the late John Coleman, of Church Cottage, Claines. He painted these for the great-great-grandchildren of Isaac John in 1998. Thanks to his wife Christine for allowing them to be reproduced here.

I thought it very fitting to support the work of the Farming Community Network (FCN) through the publication and sale of this diary. FCN is a voluntary organisation and charity that supports farmers and families within the farming community through difficult times. Further information on their work appears at the end of the book. I am sure Isaac would have approved and welcomed such support.

The diary starts on January 20th 1883, nearly one month after the death of Isaac's wife, Sarah. The first sad entry inside the cover of the diary records her clothes "put in chest".

Geoff Sansome
Claines, November 2017

THE FAMILY

Isaac John Sansome (self): Writer of this diary. Isaac John was born 9th November 1841 at Green Lane, Claines (near Astwood) as the only child of Isaac and Hannah Sansome. He was christened at Claines Church 18th November 1841. He attended St Martin's School, Worcester. In June 1856 he was awarded a prize for proficiency in Arithmetic by Revd. H.J. Hastings on behalf of the Worcester District Prize Examination. In June 1857, aged 16, he was awarded a First Class Prize (A bible) by the Worcester Diocesan Board of Education. He farmed with his father at Rushwick Farm, Worcester, then Moat Farm, Bilford Road, Claines before jointly taking up the tenancy of Oak Farm, Claines with his father in 1876. This was made up of 36 acres at Oak Farm and 30 acres at Common Hill, Bevere. He was 41 years old at the time of writing his diary.

Sarah: Sarah Ann Sansome. The first wife of Isaac John. Born Sarah Ann Stubbs in 1850, in Aston Stone, Stafford, the daughter of James Stubbs and Elizabeth Stubbs (nee Cope). She was christened at Colwall Church, near Malvern on April 15th 1868. She married Isaac John at St Martin's Church (The Cornmarket, Worcester) in November 1870. At that time her father was the Publican of the Boat Inn, 26 Lowesmoor, Worcester. The marriage was by licence and was witnessed by the Parish Clerk of St Martin's and Jane Pannier, the Baker's wife from 25 Lowesmoor. Sarah died on 22nd December 1882 aged 32, from pneumonia and premature labour. She was buried in the churchyard at Claines on the 27th December.

Rose: Rosamond Elizabeth Sansome. The first child of Isaac John and Sarah. Rose was born at Rushwick Farm in 1871 and baptised at St John's, Bedwardine, Worcester. Rose was 11 years old when her mother died. Subsequent generations of the Sansome family did not know of Rose's existence until her story was eventually discovered.

Jessie: Jessie (Jess) Hannah Sansome. The second child of Isaac John and Sarah and the first Sansome to be born at Oak Farm, in 1880. Jessie was 2 ½ years at the time of her mother's death

Pater/Father: Isaac Sansome. Isaac John's father. Born in Uplyme, Devon in 1812. He married Hannah Pippen in Axmouth, Devon in 1837 and they moved to Worcester in 1839. He worked as an agricultural labourer for Richard Spooner, Esq, MP, of Brickfields Farm, then farmed at Rushwick Farm, Worcester; Moat Farm, Astwood (The Perdiswell Estate) and retired to Northwick. He was aged 71 in 1883.

Mater: Hannah Sansome, Isaac John's mother. Born Hannah Pippen in Axmouth, Devon in 1810. Aged 72 in 1883.

Ann: Annie Greensill. Annie was the cousin of Sarah Ann, Isaac's first wife. She was born in 1856 at Dunston, Staffordshire, the daughter of Henry Greensill and Mary Ann (nee Cope). Her father Henry Greensill, a farmer, died before 1861. In 1868 her mother re-married a widower Edward Evans, a pudler in an Iron Works at Bilston, Wolverhampton. Annie, her brother James, and sister Clara were living with the Evans in 1871. By 1881 Annie was living at The Wheats Farm, Coppenhall, Staffordshire. She was living with her Uncle and Aunt, Joseph and Ann Cope, described on the census as an adopted niece. James was also the Uncle of Isaac John's first wife. Annie, aged 27, came to housekeep for Isaac John after his wife's death.

Plan 1. Oak Farm 1883

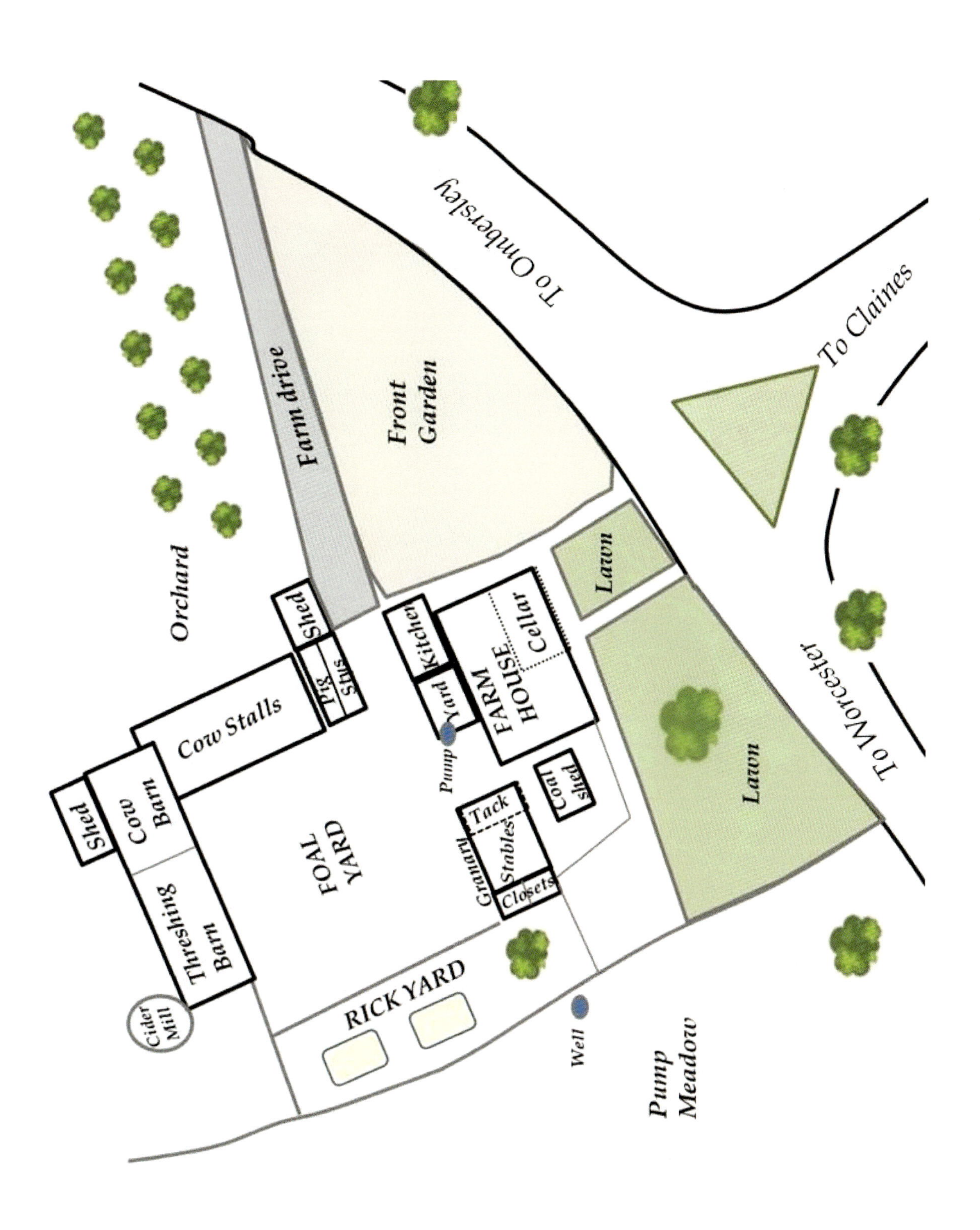

Map 1. Oak Farm 1883

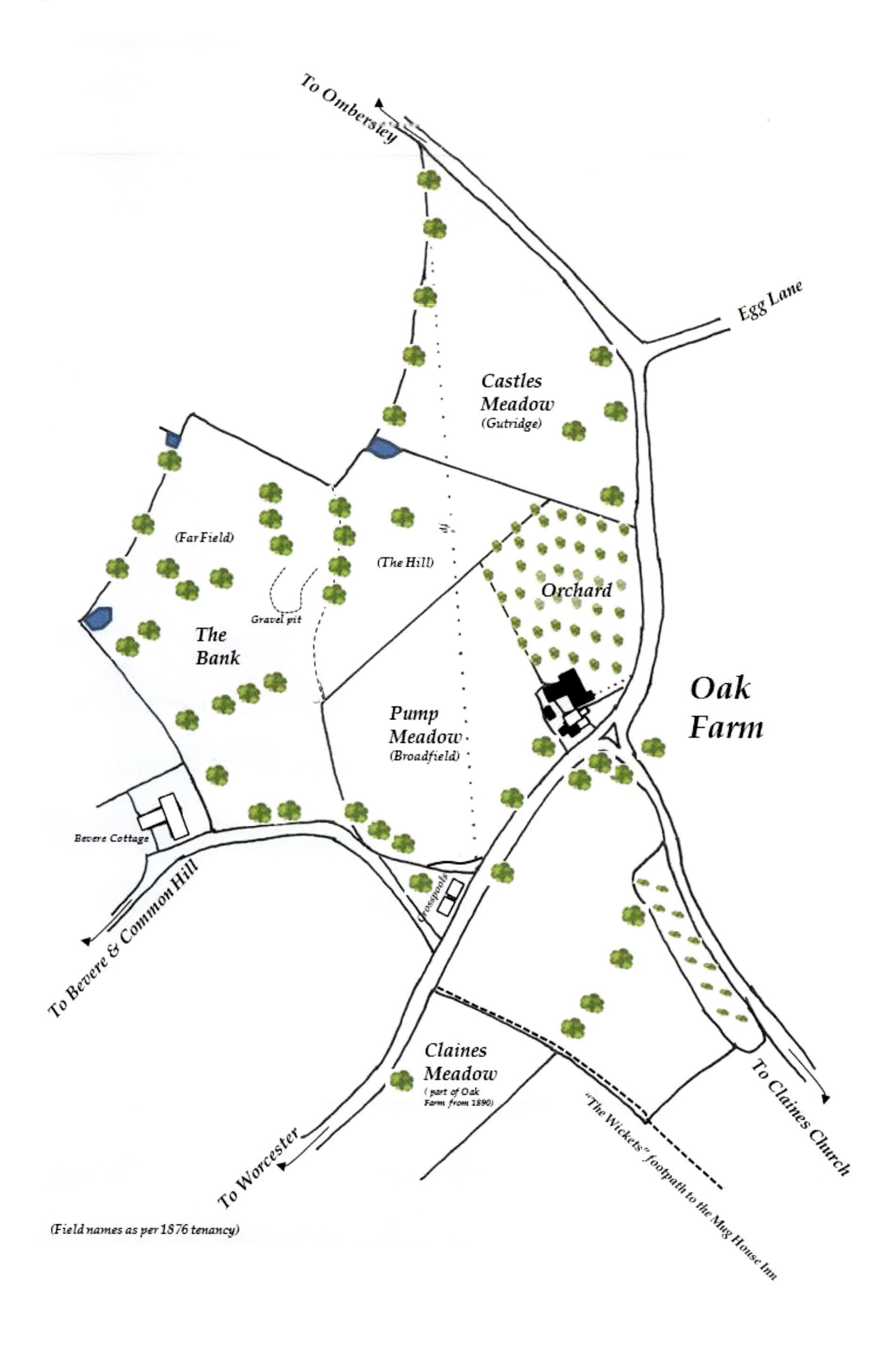

Map 2. Isaac John's Worcestershire in 1883

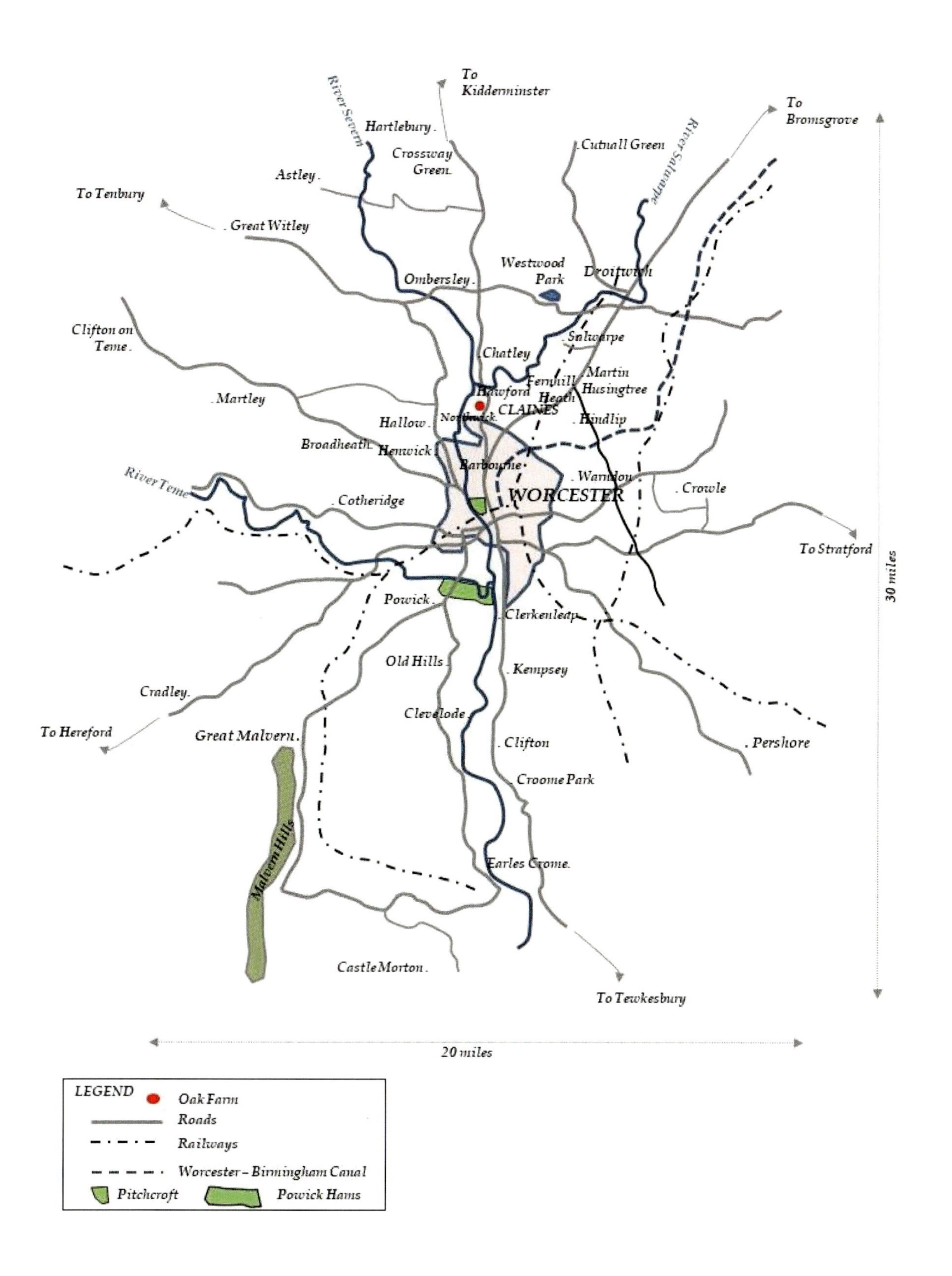

Sarah clothes put in chest

7 under bodies
7 Chemises
15 pair drawers
6 night dresses
1 short same
6 white peticoats
4 aprons
2 cloth jackets
1 silk same
2 flannel peticoats
1 sunshade
2 pairs stays
1 riding habit
1 silk dresses
1 Irish poplin dress
2 black same
1 brown same
4 print same
3 muslin same
14 collars
1 set black beads

JANUARY

One month on from his wife's death, Isaac starts his diary on 20th January. He buys himself a new coat and sets about settling his bills and receiving monies due to him for hay sales and horse hire.

The winter routines of feeding, mucking out, chaff cutting, hedge trimming and farm repairs continue, but with time found to attend two farm sales. Isaac delivers hay to his local customers and regularly travels into Worcester for cattle feed and groceries, as well as to put money into the Old Bank.

The weather turns wet and floods from the River Severn threaten his fields at Common Hill. He settles his bill for his daughter Rose's schooling. The month doesn't end well with the birth of a still born young colt.

Jan 20th

New frock coat £2-10-0 off Hansher.[1] Bowden[2] paid for 5 cwt[3] hay had yesterday, £1 paid on day, 25s for load bean straw.[4] Paid F. Smith for meat 15s.[5] Had meat off Quarterman's[6] for keep of his heifers.[7]

Jan 21st

Church a.m.[8] Northwick p.m.[9]

Mon Jan 22nd

John[10] trimming hedges and 1 truss[11] hay to Revd Curtler.[12] Common Hill in afternoon.[13] Ann to Jeavons, I gave her 5 shillings.[14]

---

1 Michael Hamsher, tailor, draper and military outfitter, 3 St Nicholas Street, Worcester.
2 John Bowden, Carpenter, Ombersley Road, Claines.
3 Cwt = Hundredweight which is 112 pounds or 50kg, with 20 cwt in 1 ton.
4 This would be the stems and leaves from field beans, used for animal bedding.
5 Frederick Smith, a Butcher at 45 The Tything, Worcester. His widowed Mother ran the Butchers, employing three men. The family previously farmed 300 acres at Hindlip Court Farm. Often referred to as F.S in the diary.
6 John Quarterman, a Butcher of 53 Droitwich Road, Claines.
7 Keep = Isaac would be "keeping" the heifers on a contract rearing/feeding basis and the meat was payment in kind.
8 St John Baptist, Claines. See photographs.
9 References to Northwick are assumed to mean his parents in Northwick Terrace.
10 John Shuck the farm labourer.
11 A truss was a bundle of hay weighing 56lb, which is half a hundredweight.
12 The Reverend Thomas Gale Curtler, eldest son of TGC Curtler. (1826-1891). Rector of Doverdale (1851-1857) Vicar and Patron of St Stephen's, Barbourne from 1861. Aged 57. Lived at Bevere Knoll.

Tues Jan 23rd

John raking thorns at hedge. To sale at Earls Court.[15]

Weds Jan 24th

John at cows, self to sale at Clifton on Teme. Linny[16] came home from Rev Curtler, had her 19 days on the 5th January. Harris[17] borrowed cart mare, to do hedging in return.

Thurs Jan 25th

John taking 5 cwt hay to Baldwin[18] making total half ton of same. Self half ton hay to Revd Curtler for £2-5-0. Fetching 30 bushels grains[19] from Vinegar works[20] and 1 bushel small corn and 1 cwt bran off Holthams.[21] Sold Mr Morton three small calves for £17-15-0.[22]

Fri January 26th

John spreading manure at Common Hill. At cows. Self to Northwick in afternoon.

Saturday Jan 27th.

Mr Morton paid £17-10-0 for three small yearling calves. Very wet. John cutting chaff.[23] Self to Common Hill to remove old Grey from lower meadows.[24] To Worcester in afternoon. Paid Miss Stearman[25] 25s 6d for Rose schooling and sent £1 to Aunt Cross for same.[26] Bought grocery and new bass broom.

Sunday Jan 28th

Church a.m. Self at cows.

Monday January 29th

John at cows washing trap and fetching old grey up. To Worcester paid 15s trap tack.[27] Put in O.B £15-0-0.[28] Old Grey mare took bad. 1 peck[29] potatoes off old Presdee 1s.[30]

---

13 The 36 acres of land at Common Hill, Bevere, which Isaac farmed.
14 Thomas Jevons, of Cumberland Street, a retired plumber and painter.
15 Earls Court was a farm in Dines Green, St John's, Worcester.
16 Linny = a horse.
17 William Harris, Market Gardener. No 8 Corn Meadow.
18 Henry Baldwin, Market Gardener, Ombersley Road.
19 Grains = Malt Vinegar is made from barley grains, which are germinated and then brewed into a vinegar beer. The spent grains are a useful animal feed.
20 The Vinegar Works. Hill, Evans & Co Ltd, vinegar manufacturers based in Worcester. Founded in 1830 and at one time the world's largest producer of vinegar, the works closed in 1965. Their buildings were in Lowesmoor and now house the Territorial Army.
21 Holthams, the farm seeds and feeds shop that stood in the corner of the Cornmarket in Worcester, see photographs.
22 Mr James Morton, a Gentleman of Hadley, near Ombersley.
23 Chaff; lengths of straw cut into short pieces and mixed into feed for horses and cattle. Cut with a chaff cutter, powered by hand or by horse gear.
24 The old grey mare horse.
25 Fanny Rose Stearman, a Governess of a Ladies boarding and day school, Ivy House, Sansome Walk, Worcester.
26 Sarah Cross of Stretton, Cannock. Born in Penkridge 1827 married to George Cross, a farmer, related to Isaac's first wife through the Cope family.
27 Tack = horse harness.
28 OB = "Old Bank". Worcester Old Bank was founded in 1772 by Berwick & Company, which flourished in Worcester for more than a century, until 1906, when it was taken over by the Capital & Counties Bank Ltd, which was in turn taken over by Lloyds Bank in 1918. In 1883 it was "Berwick, Lechmere, Isaac, Isaac and Martin".
29 A peck is a dry volume measure and is equal to 8 imperial quarts (2 imperial gallons), or one-fourth of a bushel.
30 William Presdee aged 84 of Crosspools Cottage. Next to Oak Farm.

Tuesday January 30th

John to Worcester after Robinson to Old Grey, drawed dead colt.[31] 1 pot[32] potatoes off Newey.[33] Mr Barnard[34] paid £3-17-0. John cutting chaff, self fetching 2 new gate posts from TGC.[35] Marshall putting up gate in Orchard against road.[36]

Wednesday January 31st

Marshall at gates. John at cows. Self to Whites[37] Fernhill Heath and paid him 13s for bill. To Hindlip and for ride with F.S at night.

---

31 Robinson, delivered a stillborn colt. Possibly Joseph Robinson, farm bailiff to RP Hill, Esq of Brickfields Farm, Claines.

32 Pot = Pottle, which was a volume equivalent to ½ gallon.

33 James Newey, a market gardener of 6 acres at Northwick.

34 Frederick William Barnard, a dairyman's son from St Oswald's Walk, Worcester, who moved to Hill Cottage, Bevere. He regularly bought Isaac's milk.

35 TGC: Thomas Gale Curtler, JP, Landowner of the Bevere Estate and Landlord of Oak Farm. Of Bevere House, Claines, aged 86.

36 George Marshall, an agricultural labourer of Ombersley Road, Claines.

37 Henry White, a Blacksmith of 1 Droitwich Road, Fernhill Heath (Now The Old Smithy 218 Droitwich Road).

FEBRUARY

As gate and fencing repairs continue, a visit from the vet to a sick cow is required. The first hint of spring arriving sees the grass fields being chain harrowed. Isaac's cows are sent off around the parish to various bulls to be served. Brewer's grains for cattle feed are bought and collected regularly from the Vinegar Works at Worcester and other feeds from Holtham's in the Corn Market. Isaac sends out cider to his father at Northwick. The weather turns wet and the land at Common Hill is flooded for the sixth time this winter.

Ann, Isaac's late wife's cousin, visits to help Isaac keep house and take care of his daughters Rose and Jessie and Isaac settles her wages. Isaac visits local friends and neighbouring farmers with his butcher friend Fred Smith and occasionally takes his mother on his visits.

Isaac rides out with the Worcestershire Hunt at Ombersley and Hindlip as well as hacking out 12 miles to the Old Hills to ride with them there. His young colt, Prince, is broken into harness for the first time.

Thurs Feb 1st

John fetching 30 bushels of grains from Vinegar works and grocery. Self to Miles[38] in afternoon with F.S.

Friday Feb 2nd

John unloading grains and Marshall ½ day at gates. To Common Hill in morning. Gave Rev TGC[39] man 2s cleaning harness. Mater up.[40]

Satur Feb 3rd

Marshal at gates going into bank. John at cows, self to Worcester for Vet to red half bred Hereford cow, stanked, gave same 3 drinks. To Worcester in afternoon. Paid F.S 11/4 for meat.

Sun Feb 4th

Church in morning, Northwick afternoon. Sent John to Worcester to stop Vet coming to cow at night.

---

38 Joseph Miles of Moat Farm Astwood, Bilford Road, Claines, where Isaac's father used to farm, see photographs.
39 Thomas Gale Curtler's coachman, Joseph Smith of Bevere.
40 "Up" presumably means visiting Oak Farm.

Mon Feb 5th

To Worcester to meet Ann.[41] ½ cwt bran and 4 bushel oats, 3/6 off Holthams. Paid same 19/9 up until Christmas. Revd Curtler had nag mare to go to Worcester.[42]

Tues Feb 6th

To Whites with pump handle repair.[43] John chain harrowing in bank.[44] To Mr Day's[45] in afternoon with Mater. Bought white cow off same £15-10-0 not calved yet. Sold F.S cow that was bad for £16-10-0. Gave £13-15-0 for same, profit £2-15-0.

Wednesday Feb 7th

At cows and took Annie to Mr Holland's.[46] John chain harrowing.

Thurs Feb 8th

Paid Adams[47] blacksmith £2-0-0. John to Worcester after 30 bushel grains from Vinegar works 15s. Paid F.S for last week's meat £7-10-1/2. Bought grocery 7s-5 ½. New locks and key from Halls 1s.[48] Took fathers cows to Denovans[49] but was not beasting.[50]

Fri Feb 9th

John at cows and self taking heifer (strawberry) to Mr Denovan's bull. Hunting at Ombersley.[51]

Saturday Feb 10th

John at cows. F Smith sent for half bred Hereford cow sold for £16-10s. Mr Barnard paid £3-18-9. To Worcester in afternoon. Paid Ann £1-1-0 being amount of wages due to her up to Feb 7th 1883. Gone home till next Wednesday.[52] Paid J Newey 5s for pot potatoes.

Sunday Feb 11th

At cows. 1 truss hay to F. Smith, paid same time.

Monday Feb 12th

John at cows, wet day. 6th flood on Common Hill. To Worcester put in OB £3.

Tuesday Feb 13th

John at cows. T.G.Curtler Esq had cart mare. Sale at Clevelode. 1 cwt sharps and 1 bushel fair corn 4/6 off Holthams.[53]

Wednesday Feb 14th 1883

John taking two cows to Burroughs to calve. Cutting chaff and sold F.S heifer strawberry £13. Wet.

---

41 To collect Ann (Annie) from the train.

42 Nag mare = an old female horse.

43 There were two water pumps at Oak Farm. A Thomas hand pump, over a well in the yard behind the house, and a well and pump in the "Pump Meadow".

44 Chain harrows, a square of linked iron chains, dragged behind a horse to remove dead grass and level mole hills in grassland.

45 William Day, Farmer, Tolladine, Worcester.

46 Possibly Alfred Holland, a Tailor of St Georges Lane, Claines.

47 Charles Adams, Blacksmith, 56 Ombersley Road, Claines.

48 J&F Hall, Iron Merchants and Iron mongers 1&2 The Shambles, Worcester.

49 James Denovan farmed 220 acres at Holy Claines, Farm, Hindlip Lane.

50 Beasting/bulling= in heat and ready for mating.

51 Riding with the Worcestershire Hunt.

52 To Stafford.

53 Sharpes was a brand name for a Poultry feed.

Thursday Feb 15th

John to Worcester after 30 bushel grains Vinegar works. To Mr Day's, Tolladine to see about cow. Wet.

Friday Feb 16th

John and self falling tree in garden. To Coroner Hughes[54] for 2 cwt magnum bonum potatoes for 14s.[55] To Worcester, had nag mare shoed all round at T. Abell's. [56] Dry.

Sat Feb 17th

John at cows, cutting chaff. Mowing lawn. Self to Worcester in afternoon. Had £10 off F. Smith and paid M. Day for white cow, Tolladine £15-10-0. Paid F. Smith 10s 6 for meat and grocer 2s 6. Drawed £5 out of Old Bank.

Sunday Feb 18th

To Common Hill with F.S in morning. Miss Jowett here afternoon.[57]

Monday Feb 19th

John fetching white cow from Mr Day at Tolladine. Hunting at Hindlip. J. Tearne finished cropping hedge between pump meadow and bank at 1 ½ pearch.[58] Drawed 3s on same.[59]

Tuesday Feb 20th

J. Tearne carting to bank. John at cows. I and Tustin[60] began breaking Prince to harness, had him shoed all round.[61] Paid T. Able 6s for same and nag mare. To Worcester in afternoon, rode Prince home.

Wed Feb 21st

J.Tearne carting to bank, drawed 1 shilling. John at cows. Self and strawberry heifer bulling to Barnards. With Ulmnschneider to Northwick at night. [62]

Thurs Feb 22nd

John to Worcester after 30 bushel grains. Had old grey mare shoed all round at Adams. To sale at Clerkenleap with F.S. J. Tearne not at work.

Fri Feb 23rd

John carting soil from Turnpike road.[63] J. Tearne no at work. F. Smith sent up 4 pigs bought at Clerkenleap sale at 30/6 each. Ride on nag mare in afternoon.

---

54 William S.P Hughes, of Northwick Hall, County Coroner, farming 185 Acres employing 13 men, 4 boys, 3 women.
55 Magnum Bonum (1876) A potato variety bred by James Clarks (1825-1890, Potato Breeder) his first commercial success. It was trialled at Stoke Newington, and from there it was passed to Suttons who bought the variety and introduced it in 1876. It was an excellent flavoured floury potato, popular until the turn of the century, peaking in 1890. After that it degenerated and became susceptible to blight. It was used in the breeding of King Edward, Arran Victory and Arran Banner. (Nigel Deacon, Diversity website).
56 Thomas Abel, of 24 Bransford Road, Worcester, a shoeing smith (Blacksmith).
57 Jane Jowett of The Laundry, Pershore Lane, Martin Hussingtree.
58 Perch= a length of 5 ½ yards.
59 Drew money owed, before the final payment.
60 John Tustin, farmer of 100 acres, Ronkswood and well known horse breaking family.
61 Putting Prince, a young colt, into harness for the first time.
62 Charls Ulmnschneider, Landlord of the Five Ways Inn, Angel Place, Worcester.
63 The Turnpike is the Roundhouse (Toll house) at the junction of the A449 and the A38 by Gheluvelt Park, Barbourne. The soil was possibly as a result of the construction of the Worcester Tramways Trust company which had a horse drawn tram operating to Barbourne, which started operating in 1881. See photographs.

Sat Feb 24th

John carting soil off road. Paid Colley[64] 5s for hay cutting due to him about 3 weeks before.[65] Paid J. Tearne for work done and hedge trimming. To Worcester. Put in OB £3. Paid F. Smith for meat 7s.

Sun Feb 25th

John at Church, self at cows. F.S bought up 1/2 cwt bean meal for pigs. T. Cook sent for old mare.[66]

Mon Feb 26th

J. Tearne 1 day fetching load night soil from Northwick and cleaning up yard.[67] John help at same and cutting chaff. Self to Worcester after 2cwt sharps 15s from Holthams for pigs and 1 bushel oats for Linny. 2s-11d. Gave J Tearne 1s. Prince put in single harness for first time.[68]

Tues Feb 27th

Colley cutting 2 tons 9cwt hay from Northwick at home. John carrying in hay and at cows. Hunting Old Hills.[69]

Wed Feb 28th

John chain harrowing. J.Tearne ¾ day spreading manure and drawed 6d. John and self taking 28 gallons cider to Northwick and to Burrows, Broadheath to see cows, not calved yet. Left cart at F.S to bring grains tomorrow.

---

64 Henry Colley, labourer, 5 Lowertown, Claines.

65 Hay cutting refers to cutting hay out of the hay rick using a large semi-circular knife, see photographs.

66 Thomas Cook, a horse dealer of Bransford Road, Worcester.

67 Night soil = Human excrement collected at night from buckets, cesspools, and privies and sometimes used as manure.

68 As opposed to double harness where two horses would be harnessed to pull one load.

69 At Callow End, near Malvern, about a 12 mile ride from Oak Farm.

MARCH

Hedging and chain harrowing continues, with gaps in the hedges being filled by thorns and trimmings to make them stock proof before turning out the cattle. With some winter still to go Isaac orders coal and buys in swedes for the cattle.

Isaac pays the numerous tradesmen and labourers who are working for him and keeps a reckoning of his meat account which involves barter, livestock sales and meat purchase with his butcher friend. He finds time to visit the Mug House Inn and the Raven Inn with his father and other local farmer friends.

Cow and calf sales continue and Isaac regularly hires out his horses, to Reverend Curtler, the Landlord of the New Inn and others.

An early Easter arrives at the end of the month with Isaac attending the Good Friday service at Claines Church, as well as the annual Church Vestry meeting on Easter Monday.

March 1st

F. Smith bought up 15 bushels of grains from Vinegar works, kept 1 half himself. John chain harrowing in Orchard, self in garden. Mater up in afternoon. To Mug with F.S.[70]

Friday March 2nd

J.Tearne ¾ day spreading manure, drawed 1s. John and self fetching 18 cwt coal from Fernhill Heath 11/4.[71] To Worcester to see Prince in harness.

Sat March 3rd

John taking half ton hay to Revd Curtler. John Tearne ¾ day Common Hill hillockin.[72] Paid Colley 12s for cutting 2 tons 9 cwt hay on Tuesday. Paid J. Tearne 6/6 for work done in the week, Paid F.S for meat and with 5 cwt hay 6s, sent £1 to Stafford.[73] White cow at Burrows, calved heifer calf.

---

70 The Mug House Inn, Claines Churchyard, see photographs.

71 From the coal yard at Fernhill Heath Railway station. The original station was opened on 18 February 1852; at first named Fearnall Heath, it was renamed Fernhill Heath on 1 July 1883. It closed in 1965.

72 Levelling mounds and ant hills.

73 The first of many regular payments to Stafford. It is assumed that this money was sent to Joseph and Ann Cope, where Ann was living before housekeeping. It appears that Ann travelled regularly back there, with Isaac's children.

Sun March 4th

To Common Hill and to Burrows to see white cow with F.S.

Monday March 5th

J. Tearne 1 day Common Hill ditch, drawed 6d. John chain harrowing Common Hill. To Reece's after sitting hen eggs, sat same.[74] F. Smith sent two heifers to tack.[75]

Tues March 6th

Old Bill began gapping Common Hill, drawed 1s. John taking half ton hay to Major Lavie.[76] J. Tearne at ditch Common Hill. To sale, Astley Pound with F.S.

Weds March 7th [77]

Old Bill gapping Common Hill. J. Tearne and John and self hauling thorns to gaps at Common Hill.[78] Cutting chaff. Self to Worcester with 5 cwt hay for F. Smith. Paid for in meat account. I took lame pig to same. Bought back 1 sack of Oats off same. Old Bill had 1s, Jack had 1s 6d.

Thurs March 8th

John and self fetching white cow and calf from Burrows and F.S sent up Hereford cow for her milk. To Worcester in afternoon for 30 bushels grains and 2 cwts sharps from Holthams. Called at Northwick at night. Whatmore[79] had paid for 1 truss of hay. Old Bill drawed 1s on gapping at Common Hill.

Fri March 9th

John at cows and self. Rev Curtler paid for 1 ton hay £4-0-0 and £5 for use of Mare. F. Smith and M. Wall here to look at heifers. [80] Sold red heifer 2 years old had off Denovan to same for £16. Cutting chaff and sold Rudge2 trusses hay 4s. [81] Paid J.Tearne for work 6s.

Saturday March 10th

Put in OB £9. Whatmore had old mare. Paid Tustin for Prince harness £1. To Worcester in afternoon on Linny. Paid Old Bill for hedging 2s 6d. Gave John towards wages £1.

Sunday March 11th

John at cows and self. Mr Burrows over in afternoon.

Monday March 12th

Trimming hedges, cutting chaff. John began turning bury manure in yard.[82] Whatmore had old mare part of day. Ride to Adams[83] with F.S in afternoon and to Fernhill Heath.

Tues March 13th

At cows, John turning bury manure in yard. Paid Smith[84] Coroner 14s for potatoes. To Burrows in afternoon after bean straw. Old Bill drawed 1s on hedging. Caruthers had old mare. [85]

---

74 To put fertile eggs underneath a broody hen to hatch.
75 On "tack" is a term where an animal is housed, grazed or fed by another who is paid for that on a daily basis.
76 Major Ernest Lavie, JP of Bevere Cottage, formerly 8th (The Kings) Regiment of Foot.
77 Jessie's third birthday.
78 Repairing holes in hedges with small branches and bundles of hedge trimmings and thorns.
79 William Whatmore of 1 Tinkers Cross Cottages, Claines, a Thatcher by trade.
80 William Wall, farmer, Ladywood, Martin Hussingtree.
81 Henry Rudge, Milkman of 4 Crown Lane, Claines. (now Crown Street, St Stephen's)
82 A "bury" was a term for a large mound or pile, e.g. Muck bury.
83 Probably George Adams of the Lansdown Inn, Grocer, Baker and Publican.
84 William Smith, the labourer for William SP Hughes, of Northwick Hall, County Coroner

Wednes March 14th

John at cows and bury in yard. Cutting chaff. Whatmore had old mare. To Common Hill in afternoon. Old Bill drawed 1s on hedge, making total of 8s 6d.

Thurs March 15th

John to Worcester after 30 bushel grains. Jack Tearne 1 day turning bury and drawed 1s. Old Bill drawed 1s making total of 9s 6d on hedging. To Worcester in afternoon for ride on Linny. 1 ton swedes off Mr Webster. [86]

Fri March 16th

J.Tearne ¾ day turning manure and cleaning ditch. John at Cows. Self to sale at Castlemorton. 5 heifers off F. Smith went to T. Wall, charge 10s for keep of same.

Satur March 17th

Old Bill finished hedging Common Hill. Paid him 5/6 making total of 15s. Paid J.Tearne 4/6 making total of 6s for work done in week. J.Tearne hawling manure from ditch to orchard. John at cows. Self to Worcester, had Linny shoed all round T. Abells. Paid 3.0 same time. Had calf off Mr Webster for £2-10-0.

Sunday Mar 18th

Church in a.m. Northwick p.m. At cows.

Mon Mar 19th

John at cows. Self to Worcester for grocery and cutting chaff. Prince hurt leg, self bathing him. Whatmore paid 8s for use of grey mare and 1 truss hay 1s 9d.

Tue Mar 20th

John at cows and self to Worcester to try for grains. Cleaning bridle.

Wednes March 21st

J. Tearne 1 day cleaning closet,[87] spreading manure and hawling soil from Turnpike Road. John at cows. Self hunting at Henwick. J.Tearne drawed 1s.

Thurs Mar 22nd

J.Tearne 1 day digging in garden. John fetching 12 boltens of straw from Day and 1 ton Mangolds off Webster. [88] To Worcester afternoon. 2 cwt sharpes and 1 bushel Ind corn 4/6 off Holthams. [89] Paid J.Tearne 3s.

Fri March 23rd

Good Friday. John at cows and self to Church at night.

Sat March 24th

J.Tearne 1 day, Common Hill carting and helping cut chaff. Self to Worcester. Paid F.S £1-0-0 meat. New hat 6/6. Brought Cupey home. Paid grocery 8s. New lace gown at Hall 3/6.

---

85 Thomas Carruthers, Landlord of the New Inn, Ombersley Road, Claines.

86 Anthony Webster, Farmer, Church House, Cornmeadow Lane, Claines. Swedes for cattle feed.

87 The outside toilet.

88 Bolting was to use a "batten" (the quantity of straw from two sheaves of wheat). These were used for covering ricks but also when half threshed straw, was used for cattle feed.

89 Ind=Indian corn. This was maize grains, often of varying colours, for poultry or animal feed.

Sunday Mar 25th

To Northwick.

Monday Mar 26th

John Common Hill spreading manure and self at cows. Vestry meeting.[90] To Old Hills in afternoon. Sold T.Cook Prince £16-10s.

Tues March 27th

John hawling hedge brushings.[91] Self at cows. Barnard paid £3-0-0. 1 cwt hay to Rudge 4s.

Wednes Mar 28th

John to Adams, had wheels of cart looked to. 30 bushels grain from Vinegar works. Put in OB £2-10-10. Whatmore had 2 trusses hay 4s. To Raven with Den.[92]

Thurs Mar 29th

John fetching 14 cwt coal from Fernhill heath. John 1 day taking faggots[93] to Fr.[94] And bringing part load sawdust off Adams. John fetch roll from F. Smith's with old grey mare. [95] Paid J.Tearne 2s for work done in week.

Fri Mar 30th

To station with Ann, brought back grocery. Self and John in afternoon taking back F.S roll. Had ½ cwt calf meal off Holthams. 1 ½ cwt hay to Mr Whinfield. [96]

Satur Mar 31st

John chain harrowing Common Hill, cutting chaff. Self to see Colley about cutting up hay. To Worcester in afternoon. Paid Bill Miller £3-13-0.[97] Planted lettuce in garden.

---

90 The Vestry meeting was the annual church meeting to elect churchwardens, now held as part of the annual Parochial Church Council meeting.
91 Hedge cuttings.
92 The Raven Inn, public house in Droitwich Road, Claines (with Denovan).
93 Bundles of sticks for burning as fuel.
94 Fr = Father.
95 Presumably a flat roll for grassland.
96 Edward Wrey Whinfield (1826-1902), of Severn Grange, Claines. He was a personal friend of the composer Edward Elgar and a significant figure in the Worcester Music scene, aged 57. He described himself as an "annuitant".
97 Henry Bill of Mildenham Mill on the River Salwarpe in Claines (known locally today as Bill's Mill), see photographs.

APRIL

The apple and pear trees in the farm orchard receive a spring pruning. Isaac's hay sales to local Gentlemen continue and the Reverend Curtler makes frequent use of Isaac's nag mare for his journeys down to his parish church at Barbourne, for which pays five shillings a day. Mr Jeavons, a family friend sickens and Ann is sent to see him before he dies.

Isaac supplies the Landlord of the Five Ways Inn, Worcester, with cider and buys back beer in return. He buys in mangolds for cattle feed as the cows are still in and the grass is sparse. Stones are picked from the Orchard and the other fields and Isaac pays his half yearly rent of £133 to the Landlord Thomas Gale Curtler.

Prince the colt is sold and the cows manage to break the shafts of Isaac's horse trap.

Some drainage work is undertaken around the house and buildings, potatoes planted in the garden and carrot seed purchased by Ann. Isaac plans ahead for this year's hay making by purchasing another mare for his hay wagon. The last of the hay is cut from one of his hay ricks.

Sun April 1st

Self at cows, to Northwick at night

Mon April 2nd

John pruning trees in orchard. T.G Curtler had old mare 1 day 5s. 4 bushels of oats off Bill 3s 4d bushel. Sold 3 ducks to Aaron 8s 6d. [98] To Worcester fair.[99] Mr Whinfield had ½ cwt hay.

Tues April 3rd

John at cows. Self taking 5 cwt hay to Mr Whinfield. Colley cutting 1 ton 15 cwt hay rick at home. Hunting Crowle and Pershore. F up all day keep house. [100] TGC paid 12/9 for corn had last October.

Wednes April 4th

To Worcester. Put in OB £4. Bought 6 draining pipes off Butler for 2/6. Called see Mr Jeavons, very bad. John taking ½ ton hay to Revd Curtler £2-5-0. Self fetching 1 ton mangolds off Mr Webster. Tomley 1 day in front garden 2s. F. Smith had 3 pigs £2-2-0 each.

---

98 Most likely Henry Aaron, Inn keeper of the Crown & Anchor, Hylton Road, Worcester
99 The regular livestock fair (market).
100 Father.

Thurs April 5th

John chain harrowing Common Hill. To Vestry meeting at church.[101] Self cutting chaff. John taking 5cwt hay to Mr Whinfield. Sent Ann to see Mr Jeavons.

Fri April 6th

John chain harrowing in Bank. Self taking 57 gallons cider to Ulmnschneider at 9 ½d. Paid same time. Aaron paid 8s 6d for 3 ducks. To Worcester in afternoon with carcase sheep for F.S. Paid Hillman's[102] man for cows to bull 2s 6d, and paid Colley hay cutting 6s. Bought back grocery. Mr Jeavons deceased.[103]

Sat April 7th

John to Newey's after 200 plants. Self at cows. T.G Curtler had old grey mare 1 day. Mr Baldwin paid £1 towards half ton hay leaving due 16s. Sent £1 to Stafford. Put in OB £7. To Worcester to see T.Cook.

Sun April 8th

Church a.m.

Monday April 9th

John at Common Hill cleaning up meadows. Self put drain down in road way. Adams had pump handle repair. Self cutting chaff. Mrs Stock stone picking in bank, 1 day.[104]

Tues April 10th

John at cows, chain harrowing in orchard. Mrs Stock ½ day.

Wednes April 11th

John chain harrowing. Mrs Stock stone picking. To Worcester after 30 bushels grains, Vinegar works. Adams put pin in shaft spring cart.[105] T.Cook paid £16-10-0 for colt Prince.

Thurs April 12th

Mrs Stock stone picking orchard. John at cows. Self to Worcester with F.S at night. F.S paid £60 for 3 heifers. Self at Common Hill meadows. Mr Barnard borrowed trap and harness.

Fri April 13th

Mrs Stock Common Hill 1 day for F.S. John getting up hedge trimmings, self mowing lawn. F. Smith sent up Alderney cow and calf.[106] 9 gallons beer off Ulmnschneider at 6d.

Sat April 14th

John at cows. Self cutting chaff. Mrs Stock 1 day stone picking. Paid her 6s for work done in week. To Worcester. Paid Adams for repair pump handle 4s and paid Mr Curtler rent, £132.[107] Bought grocery and paid Ulmnschneider 4/6 for 9 gallons beer. Mr Barnard paid £2 towards milk money.

---

101 This second Vestry meeting was most likely the annual Church meeting, following the earlier meeting which elected Church wardens.
102 Most likely Thomas Hillman, a farmer of Hill House, Great Witley.
103 Thomas Jevons, aged 79.
104 Stone picking= as it says, picking large stones from fields to prevent damage to machinery and make them easier to manage.
105 A two wheeled horse cart with leaf springs.
106 The Alderney was a breed of dairy cattle originating from the British Channel Island of Alderney, though no longer found on the island. The pure breed is now extinct, though hybrids still exist.
107 This rent would have been due on Lady Day, 25th March.

Sun April 15th

To J. Miles with F.S. Self at cows.

Mon April 16th

John digging down to drain from cellar. To Worcester, had 2 new hind shoes put on mare. ½ cwt bean meal off Bill.

Tuesday April 17th

Tommy Hughes ¾ day in garden.[108] John chain harrowing in bank. Mr Curtler up to see drain.

Wednes April 18th

Tommy Hughes ¾ digging fresh drain. John at cows and turning bury in Orchard. Self off fetching 1 ton mangolds off Webster. T.G Curtler Esq had old grey mare 1 day making total had of 3 days at 5s day. To Worcester after ½ cwt Patterson's Victoria potatoes[109] off Packman, [110] 4s for planting. 1 draining pipe off Butler. Barnard brought back trap and harness. Ann bought carrot seed from Worcester. F up all day measuring fence with Bourne.[111]

Thurs April 19th

T. Hughes at drain in orchard 1 day. Paid him 5s for work done in week. John getting up hedge brushings. Ride to Hindlip in even. Theatre at night.

Fri Apr 20th

John chain harrowing bank. Ombersley afternoon with Ulmnschneider. F.S here at night.

Sat April 21st

John fetching stored straw from Revd Curtler's. Self cutting chaff. To Worcester got grocery in afternoon. Gave Revd man 2s.

Sunday Apr 22nd

Church in a.m. and drove J Greaves and self back to Ombersley.

Mon April 23rd

To Worcester fair. John chain harrowing in bank and bought trap down Worcester in afternoon. 4 bushel of Ind corn off Bill, 4/6 bushel and ½ cwt calf 11s 6d meal off Holthams. Mater paid Cosfords bill £5-0-0.[112]

Tues April 24th

To sale at Cotheridge. Cows broke shafts of trap and bring same to Wilson, 5s.[113] John chain harrowing bank and orchard.

Wednes April 25th

John fetching 30 bushels grains from Vinegar works. F. Smith had half of same 7/6. Gave Ann 5s. Took arm chair to Pater. J. Smith up. I bought old grey for £4-10-0 to have wagon in hay making. 1

---

108 Thomas Hughes, Agricultural labourer, 5 Dilmore Lane, Fernhill Heath.

109 A potato variety created by Patterson in Scotland 1856.

110 William Packman, Fishmonger and Fruiterer, The Cornmarket, Worcester, see photographs. Isaac's son Harry, born in 1892 went on to marry William Packman's granddaughter, Leah Packman.

111 Thomas Bourne, of Camp Meadow Farm, Bevere.

112 Cosfords the Drapers, 51 London Road. Worcester.

113 John P Wilson, Farmer at the Firs Farm, Bevere.

cwt hay to Whinfield. Sent to Chambers[114] for heifer calf bought of F. Smith 50s. F.S fetched calf from Alderney.

Thurs April 26th

John cleaning meadows, Common Hill. ½ cwt hay to Whinfield. F. Smith up, sold Linny mare to Mr Ballinger for £25-10-10. Gave J. Greaves 11s of same to take colt Prince to see Fraser tomorrow.

Fri April 27th

John meadows Common Hill. To Ombersley afternoon with Prince to show Fraser. To Worcester put in OB £20. Another Alderney cow came from F.S.

Sat April 28th

John at cows. Self cutting chaff. To Worcester, paid J. Smith £2-10-0 for calf from Chambers. Paid Hale 6s 9d for grocery.[115]

Sun April 29th

Common Hill in a.m. Mater up

Mon Apr 30th

John chain harrowing in bank. Gave Raisin 5s.[116] Colley cutting 2 tons hay, 10s. Finished cutting rick. John taking 15 trusses hay to Mr Whinfield, total now had 1 ton. Self ½ ton same to Major Lavie. Mr Cook sent for colt Prince.

---

114 Chambers, farmers at Astwood Road.

115 William Hale, Grocer, 11 St Swithin Street, Worcester.

116 John Raisen, the Gardener to Whinfield's, at Severn Grange, Claines.

MAY

May day sees Isaac plant his runner beans in the house garden and the fields are shut up to allow the grass to grow for mowing later in the year. The cows and heifers are put out to graze at Common Hill and the other fields are cleared of rubbish and chain harrowed. Mr Barnard settles his monthly payment for the milk which Isaac sells to him for his milk round.

Isaac keeps note of his Heifers ready for bulling and arranges for them to go to the Bull to be served. A newly purchased calf dies the night that Isaac brings it home.

Work continues in the farm house garden with the planting of Beetroot and the longer and lighter nights allow Isaac to ride out at night, sometimes with Ann. A mysterious entry of a row of kisses appears in the diary.

Isaac attends Claines Church fete on Whit Tuesday and manages to visit the Raven Inn and Mug House with his father.

Tues May 1st

Mr T.G Curtler had old mare ½ day. John and self cleaning up rubbish at Common Hill, chain harrowing in orchard. Planting kidney beans. Laid up orchard for mowing.[117] Mr Barnard paid £4-11-0. Mr Barnards pony to tack.

Wednes May 2nd

John chain harrowing in Pump meadow. 9 gallons of beer off Ulmnschneider 4/6. 5 cwt hay to Mr Baldwin £1. F. Smith sent up nag mare.

Thurs May 3rd

John and self at Common Hill repairing fence and taking down cows and heifers to 4 acre field at Common Hill. Mr Whinfield's two horses to tack. Mr Cook sent up horse. F. Smith fetched calf from last Alderney that came. To Mug with Pater.

Fri May 4th

John raking rubbish in bank and to Fernhill Heath after 14 cwt coal, paid for two lots of same, 18s. To Common Hill a.m. and to Worcester in afternoon.

Sat May 5th

John to Adams after load sawdust. To Worcester. Put in OB £7. Bought grocery and paid T. Abell 1s 2d and Ulmnschneider for Beer 4/6. J. Smith paid £4-5-0 for old grey mare.

---

117 Laid up= excluded any livestock so as to be able to take a hay cut later in the summer.

Sun May 6th

Common Hill in morning. J. Smith sent for old grey mare.

Monday May 7th

John taking half ton hay to Rev Curtler, £2-5-0. Raking up rubbish in bank. To Worcester fair. Major Lavie paid £2-5-0 for half ton hay had last week. Paid baker £3 and Larkworthy[118] £1-13-0 and T. Abel 1/6.

Tues May 8th

John taking white cow bought off Day to Mr Denovan's bull. Raking up rubbish and chain harrowing in bank with T. Cooke's horse. To Hallow in afternoon to see J. Smith about grain straw off Mr Webster.

Wednes May 9th

John raking up rubbish in bank. Jessie came home.[119]

Thurs May 10th

John hawling of rubbish in bank. Rev Curtler paid for 1 ton hay £4-10-0. To Worcester in afternoon. Paid Holtham £6-7-0 and bought nails at Halls. Sent £1 to Stafford. To Broadheath with F.S. ½ cwt calf meal off Holthams.

Fri May 11th

John and self in front garden and hedging. To see Rev Smallwood's calf with F.S.[120] Mater up

Satur May 12th

John and self finished hedge. Self to Worcester for grocery. xxxxxxxxx[121]

Sunday May 13th

Common Hill in a.m. Northwick p.m.

Monday May 14th

To Worcester fair with Pater. John digging in garden, bought up heifer from Common Hill.

Tuesday May 15th

In garden planting beetroot. John and self to Claines Fete.[122]

Wednes May 16th

John at cows and self to Barnard after harness. To Martley in afternoon with F.S. Had Wilson trap.

Thurs May 17th

John cleaning up rick yard and fetched bull for F.S from Mr Denovans. Self to Common Hill.

---

118 Larkworthy the Agricultural Engineer, of Lowesmoor Iron Works, Worcester.

119 Isaac's daughter Jessie, home from Stafford.

120 William John Smallwood, Vicar of Claines 1870-1883.

121 A row of 9 "kisses".

122 Claines Church Fete, see photographs. It may seem strange to have a Fete on a Tuesday but this was Whit (sun) Tuesday and a traditional time for Fetes and Fairs.

Friday May 18th

John fetching ½ cwt bran from Bill, 3s and clearing up rick yard.[123] Drove self to Raven with Pater. To Worcester in afternoon, bought pair boots 12s 5d off Tyler. Heifer from Chambers' cow calved bull calf. Gave Ann 1 sovereign.[124]

Satur May 19th

John at cows, self looking to heifer and calf. To Worcester in afternoon to order meat.

Sun May 20th

To Common Hill in evening

Monday May 21st

John to Worcester after iron fencing. Self to fair. Bought calf off for F. Smith 34s, died same night. F.S had it to Fernhill Heath. Old Alderney went.

Tues May 22nd

John to Worcester with dead calf and brought back hay, oil cake and flour for bull which came for F. Smith.[125] To J. Miles for cider and at Paters. Butler doing drain from sink. Oats from Webster for F.S bull.

Wednes May 23rd

To Worcester auction sale with F.S. Horse not sold. John waiting for Hall's man with fencing. Not come.

Thurs May 24th

John white washing back kitchen.[126] 2 cwt bean flour from Bill for F. Smith. To Gregory's Mill[127] to look at Mare belonging to T.Cook, not of use. 1 bushel wheat off Bills.

Friday May 25th

John at cows. Big white cow bulling, bulled by small red one from Denovans. To Worcester in afternoon after grocery and to Hindlip. Had 4 pigeons.[128]

Satur May 26th

To Worcester horse sale for Revd. John repaired Bull's tie, cleaning up wood heap.[129]

Sun May 27th

To Common Hill a.m. To Malvern for drive p.m. with Mr Ulmnschneider. John at cows. To J Miles with F.S in a.m.

Monday May 28th

John to Worcester after 9 gallons beer Ulmnschneider, and in afternoon to change horse for mare. Took boots to Tyler's.[130] Halls sent up remaining fencing.

---

123 The location of the hay and wheat ricks at Oak Farm.
124 A gold coin, worth 20 shillings or £1.
125 Slabs of cake for cattle feed, the residue of Linseed when it had been pressed for oil. Broken up by a hand operated cake breaker.
126 The small outbuilding at Oak Farm, still known as the back kitchen.
127 Off Barbourne Road, still operating as a corn mill during the 1914-18 Great War.
128 Most likely "shot".
129 The tether where the farm Bull was tied up.
130 William Henry Tyler, Bootmaker, 9 St Martin's Gate, Worcester.

Tues May 29th

John cutting chaff and throwing up bury in yard, cleaning harness and self fetching spotted cow up. John drawed 1s 6d.

Wednes May 30th

John at cows. Self to Wilkes for mortar for Butler and to Butler for 10 pipes for stable drain. Butler doing ash pit. To Brickfields with notices for tenants and to OB put in £4-10-0.[131]

Thurs May 31st

John and self to J. Alfords with notice and to Droitwich with F. Smith and to Wilson about changing trap shafts and to J Miles in evening. New tar sheet off Griffiths, 6 yards by 4 yards.[132]

---

131 It is not known to what "notices for tenants refers". Possibly Isaac was acting on behalf of the owners of the Brickfields Estate?

132 Henry Griffiths, a rope and marquee manufacturer of 38 The Tything (Close to the Green Man Public House).

JUNE

Isaac orders and helps set the headstone for the grave of his late wife Sarah, in Claines churchyard. The cooking range in the back kitchen is repaired and some iron fencing is put up around the farmyard. Isaac is hauling rubble and soil from the renovations at Claines church to fill in a ditch at the farm.

Preparations are made for haymaking, with Isaac having repairs to his spring cart and hay pikes, as well as buying a new rick sheet in case of rain. He finds time to attend a visiting Circus and the Worcestershire Agricultural Society annual show. He attends the weekly livestock fair at Worcester to sell a calf.

Isaac has sold all of his own last year's hay, but still has customers wanting some. He buys in hay and sells it on to help them out. He pays back his mother some money borrowed and also sends money up to his late wife's relatives at Stafford who help look after his daughters.

He buys in beer ready for the haymakers and visits farmers in neighbouring parishes to see if he can get some cider. The mowers arrive to knock down the grass for haymaking.

Fri June 1st

John wheeling rubble to fill up ditch. To Worcester with trap for grocery and to Thornbury[133] to see about stone for Sarah[134] and trial with F.S mare with Bridges. Sent trap back to Wilsons.

Satur June 2nd

John hawling rubble with mare from Claines churchyard to fill up ditch in foal yard.[135] Self to Worcester in after, to show man the mare. Agreed with F.S to have spotted cow and £3 for mare. Thornbury drawed 1 sovereign on stone.

Sun June 3rd

At Cows and to Droitwich in afternoon with F.S.

Monday June 4th

John to Common Hill counting iron fencing and put heifers in F.S bank.[136] To London Road to see G. Crisp with club money.[137] Mr Payne sent heifer to bull.[138]

---

133 Jesse Thornbury, Stone Mason of 12 Northwick Place.

134 Grave stone for his deceased wife, Sarah.

135 Rubble associated with the Aston Webb alterations to Claines Church, completed 1886. The foal yard (as per the farm plan) would most accurately be described as a "fold yard" which is a yarded enclosure for cattle. But it was known within the family and described in the diary as the foal yard, probably a derivative of the correct name.

Tuesday June 5th

John to Fernhill Heath after ½ ton coal 7s-3d, and hawling dirt from Claines Churchyard. New spanner off HT White. F. Smith sent for big red bulls being tied up for about two weeks. Gave John 1 sovereign towards wages making total of £3-10-0. Mr Evans doing back kitchen range, paid him for same 2s 9d.

Wed June 6th

John at work in garden and cows and heifer from Webster supposed to go to bull of F. Smith from C. Herbert.[139] To Worcester in aft and had 2 new shoes put on nag and heels trimmed. Spotted cow calved. To Northwick in a.m.

Thurs June 7th

John in front garden. To Droitwich and to Burrows and circus with F. Smith. John circus at night. Revd Curtler had nag mare Polly twice to Worcester. Sent spring cart and 11 pikes to Adams for repair.[140] Paid Mater sovereign borrowed.

Fri June 8th

John work in garden. To Common Hill and to Adams about Spring cart. To Worcester in aft[141] after grocery with Carruthers, and to Wilson's after own trap. To go back again shafts not straight. 1 peck potatoes off Newey. Paid F. Smith £24 in settlement of mare, meat, spotted cows and sent 1 sovereign to Stafford.

Saturday June 9th

John in garden. Self to Worcester in afternoon, paid Tyler 2/6 for repaired boots, Amphlett for oil and cow salts.[142]

Sun June 10th

Church a.m. Northwich p.m.

Mon June 11th

John removing rails from bank Common Hill to bottom meadow. Cleaning bits.[143] Self in garden. Paid Webster £4-13-0 towards mangolds, calf and straw owing £1-10-0.

Tues June 12th

John assisting Mr Hall's man putting up fencing Common Hill. To Old Hills in afternoon for ride with Mater.

Weds June 13th

John taking Hereford cow to Mr Wall's at the ford. Self in garden, to Northwick in afternoon.

Thurs June 14th

John 1 day, Common Hill. Repaired road to rick yard. Self in garden to Worcester in aft for grocery.

---

136 This refers to land that Frederick Smith, Isaac's butcher friend had at Common Hill.
137 George Crisp, the father in law of Cosfords the tailor, of 51 and 52 London Road. The Club referred to is possibly the Ancient Order of Foresters, as Isaac's father founded the "Court of Hope" at Worcester in 1852.
138 John Payne, a farmer of 28 acres at Chatley.
139 Charles Herbert, farmer at Tapenhall Farm.
140 Hay forks.
141 Aft=afternoon.
142 Edmund Amphlett, a druggist of 8 Mealcheapen Street, Worcester.
143 Bridle bits.

Fri June 15th

John taking white cow to Common Hill to bull of F.S bought off C. Herbert. Washing trap. Self planting plants in garden, had off Raisin. Sold F. Smith spotted cow for £26 in settlement of meat, keep of heifers. Self 56 gallons of beer off Ulmnschneider at 6d.

Satur June 16th

John in garden. To Worcester in afternoon. Gave Ann 1s.

Sun June 17th

Common Hill, cows.

Monday June 18th

John in garden. Self to Worcester fair, sold calf from light heifer off Chambers to Mr Bayliss 17s/6. Half ton hay to Revd Curtler had off J. Miles for £2.

Tues June 19th

John repairing road Common Hill. To J. Miles paid £2-0-0 for half ton hay. To Northwick afternoon.

Wednes June 20th

John and self taking Chambers' heifer to bull Common Hill, no go, supposed by bull at home. John fetching celery plants from Northwick. To Gresley's with F.S in afternoon.[144] F.S sent for old Alderney.

Thurs June 21st

John at cows. Self to Agricultural show.[145]

Friday June 22nd

John washing trap, mowing nettles.[146] Self to Adams after spring cart and to Banks for rick sheet.[147] 2 heifer off F. Smith came from Droitwich. Whinfield's one horse taken away.

Satur June 23rd

2 Paynes heifers came 2nd time. John mowing nettles. To Worcester with trap for grocery.

Sun June 24th

At home. John and Ann. Ann 1s 9d.

Monday June 25th

John cutting thistles Common Hill. Took heifers out of F.S bank, spotted cow and calf went sold to F.S £26-10. T.Cook had his horse away. F. Smith took his two heifers and calf away that came on Friday.

Tues June 26th

John and self helping to put stone to grave. Paid Thornbury £5-5s for same and Revd Smallwood £1-1-0. In garden, to Northwick in afternoon. Barnard paid £4-0-0.

---

144 Robert Archibald Douglas Gresley, a magistrate and farmer of 220 acres at High Park, Droitwich, aged 89.
145 Show of the Worcestershire Agricultural Society, formed in 1830, prior to its amalgamation with Herefordshire in 1894, subsequently becoming the Three Counties Agricultural Society. *"In 1883 foot and mouth disease broke out at Ronkswood and Astwood farms in June which excluded cattle, sheep and pigs from the showground at Battenhall; a wet second day soaked the three flower tents and ruined the attendance."* (A history of Worcestershire Agriculture: R.C. Gaut 1939)
146 Two wheeled horse carriage.
147 Henry Banks rope manufacturer, 36 The Shambles, Worcester. A rick sheet was used for temporary protection of the hay rick until it was thatched.

Wednes June 27th

John mowing thistles Common Hill. Captain Castles[148] sent his heifer to bull. T. Cook sent two horses for work. To Worcester station after Rose. Put in O.B £20. Paid Webster 30s owing him.

Thurs June 28th

John hawling manure to bury in Captain Castle's meadow. To Martley to try for cider and to J. Miles to order hay.

Fri June 29th

John hawling manure. Self to Worcester in afternoon. Took Mowers from Captain Castles to Ombersley and paid for 18 gallons cider 9d to Mr Horton. [149] Phillips borrowed 1 sovereign. To Worcester with barrel for beer to Ulmnschneider and to Burroughs with F.S. Half ton hay off J. Miles to Major Lavie.

Satur June 30th

Began mowing opposite Major Lavie's. John fetching wagon from Barnard's and took same to Adams. To Worcester in afternoon bought 18lb cheese for haymaking at 5d and 2 new rakes. Whinfield's horse came.

---

148 Captain Charles Castle of Hawford House, High Sherriff (1881) and Deputy Lieutenant of Worcestershire. The meadow still bears his name.
149 John Horton a farmer of 230 acres, at Suddington, a hamlet of Ombersley.

Oak Farm lease documents

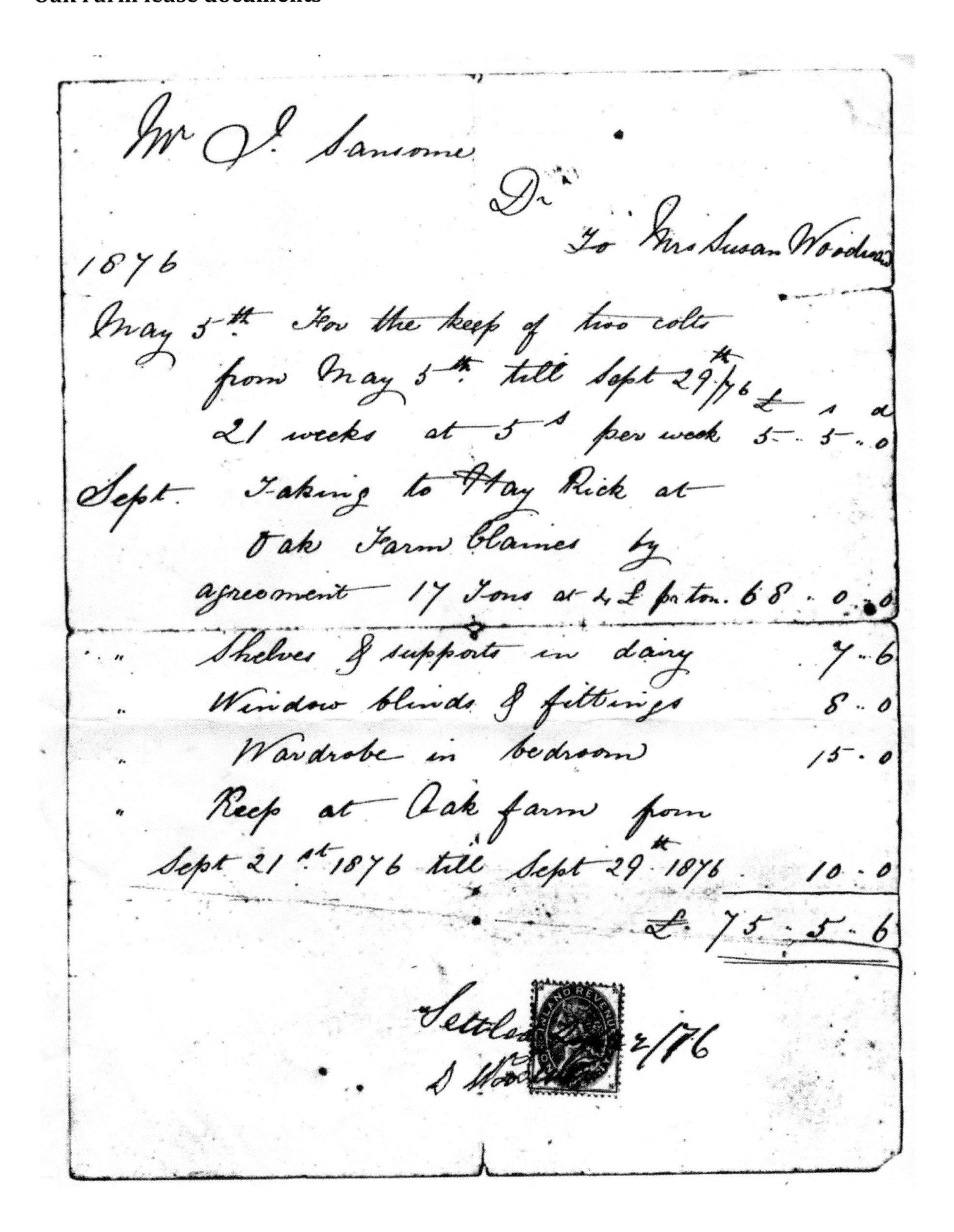

Mr I. Sansome

Dr

To Mrs Susan Woodward

1876

		£	s	d
May 5th	For the keep of two colts from May 5th till Sept 29th/76 21 weeks at 5s per week	5	5	0
Sept.	Taking to Hay Rick at Oak Farm Claimes by agreement 17 Tons at 4£ per ton.	68	0	0
"	Shelves & supports in dairy		7	6
"	Window blinds & fittings		8	0
"	Wardrobe in bedroom		15	0
"	Keep at Oak farm from Sept 21st 1876 till Sept 29th 1876		10	0
		£ 75	5	6

Settled [illegible] 2/76

S Wood[illegible]

1. Invoice for the incoming valuation at Oak Farm, September 1876, from Susan Woodward to Isaac Sansome

Dated 29th July 1876.

Thos. Gale Curtler Esqre

to

Messrs Sansome

Copy

Lease of Oak Farm

Claines.

Rent £268.

The original of this on Stamp is in my possession

T G Curtler

2. The lease agreement for Oak Farm between T.G Curtler and "Messrs Sansome" (Isaac and Isaac John) 1876

Schedule of the demised Premises.

Oak Field Farm.

	a	r	p
House Garden &c.	0	1	29
Broadfield	7	2	34
Far field	11	1	20
The Hill	4	2	33
Gritridge	8	3	2
Orchard	3	2	36
	36	2	34

Common Hill Lands.

	a	r	p
Big Meadow	9	1	2
Camp Close	4	1	4
The Hill	5	1	20
Wet Ley	5	1	28
Upper Pit	2	3	30
The Pit	3	0	16
	30	1	20
Total A.	67	0	14

3. Schedule of land at Oak Farm and Common Hill

(67 acres, 0 rods, 14 perches)

Photographs

4. Isaac John Sansome with his family at Oak Farm, 1900

Back row from left; Jessie, 19, John Isaac (Jack), 13, Annie, 15,
Front row from left; May, 9, Isaac John, 58, Harry, 7, Annie (second wife), 43, Lily (twin to May), 9

5. Moat Farm house, Bilford Road, Astwood, Claines where Isaac John farmed until 1876

6. Moat Farm

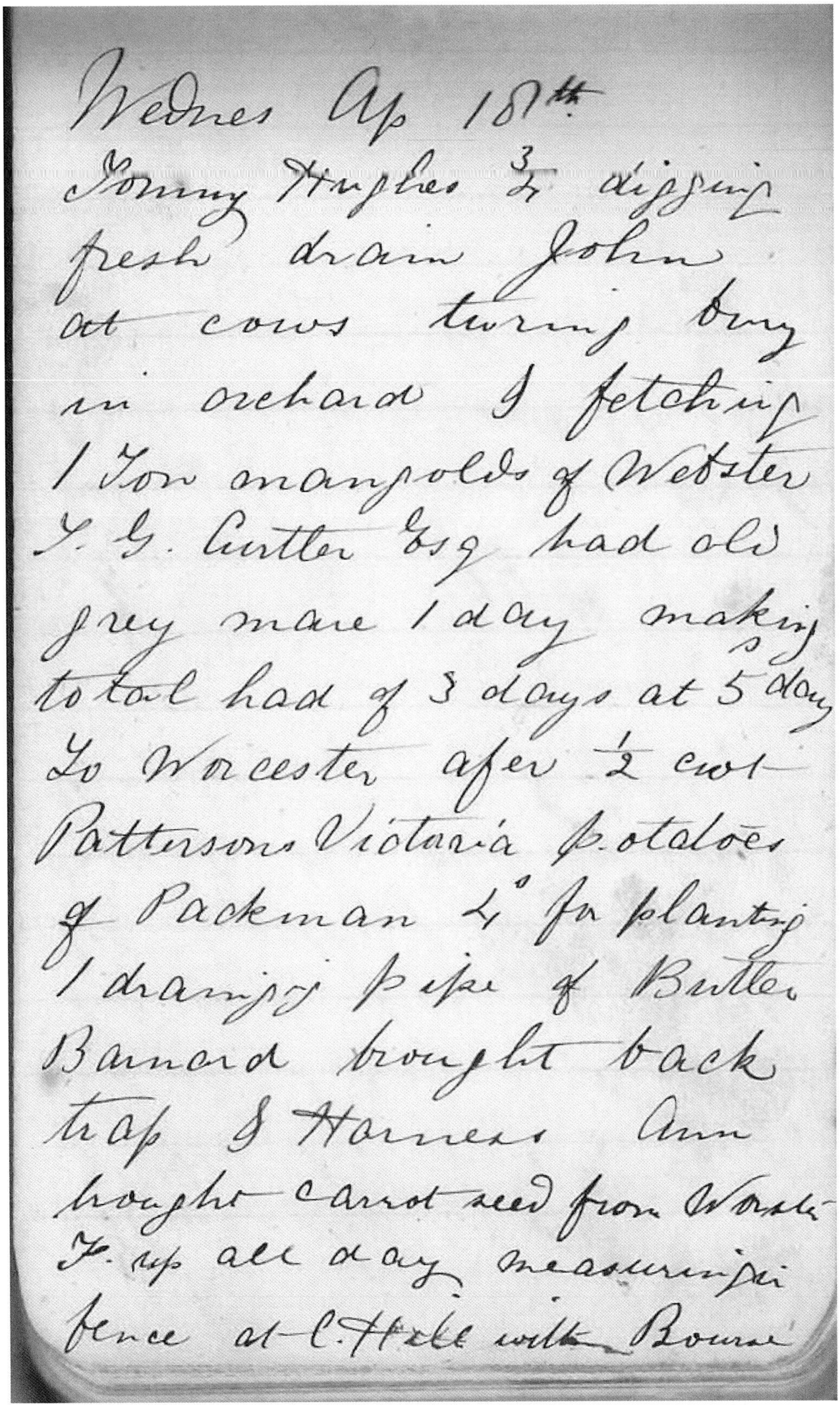

Wednes Ap 18th
Tommy Hughes 3/4 digging
fresh drain John
at cows turning dung
in orchard I fetching
1 Ton mangolds of Webster
T. G. Cutler Esq had old
grey mare 1 day making
total had of 3 days at 5s day
To Worcester after 1/2 cwt
Pattersons Victoria potatoes
of Packman 4d for planting
1 draining pipe of Butler
Barnard brought back
trap & Harness Ann
brought carrot seed from Worster
F. up all day measuring
fence at C. Hill with Bourne

7. The page for April 18th 1883, from the original diary of Isaac John Sansome

8. Cutting hay from the rick in Claines

9. Believed to be Isaac John's cattle, at Common Hill, Bevere

10. Oak Farm House in 1915

11. Isaac John, sat on a Bamford Hay rake in the Orchard at Oak Farm, with son Harry. Circa 1917

12. Isaac John with gun and dogs in the Orchard, late 1920's.

13. Claines Church, circa 1900

14. The Mug House Inn, Claines Churchyard

15. Claines Church Fete, Severn Grange 1909

16. The Turnpike, Barbourne

17. Hawford Mill, Claines. William Warner, the miller in 1883

18. Mildenham Mill (Bill's Mill) at Claines. Henry Bill, the miller in 1883

19. Holthams, the Seed and Corn Merchants in the Cornmarket at Worcester

20. The Cornmarket, Worcester, showing the shop of William Packman, Fishmonger and Fruiterer

JULY

Haymaking dominates the month. Isaac borrows wagons and a hay tedding machine from neighbours and buys in food, drink and hay rakes. He borrows some money to help him through the hay making expenses.

The mowers come and go as the wet weather regularly interrupts hay making. He takes advantage of the wet weather to attend Worcester races. Building of the hay ricks starts and this requires regular moving and erecting of the rick poles.

Isaac buys in 144 gallons of beer and 18 lbs cheese to keep the hay makers going and has to hunt them down in neighbouring parishes when he wants them to return. It's Rose's 12th birthday and she spends some time at Hindlip. Isaac attends church with Ann and a romance seems to be building as Ann is mentioned along with several kisses after her name.

Sunday July 1st

At home

Mon July 2nd

Haymaking bank. John fetching 60 galls beer and changing T. Cook's horse from Ulmnschneider and own wagon from Adams.

Tues July 3rd

Haymaking in bank, hawled load. John fetching wagon from J. Smith, Hallow.

Wednes July 4th

Haymaking in bank. Hawled 1 load. Wet aft.

Thurs July 5th

Wet. Took nag had 4 shoes on T. Abell's. To Worcester races, John drawed 2s.

Fri July 6th

White cow to Mr Dean's Bull,[150] haymaking in Bank

Sat July 7th

Haymaking. To Worcester in aft for grocery.

---

150 George Dean, Gardener and Bailiff of the Perdiswell Estate.

Sunday July 8th

F.S Red bull took to Common Hill. After mowers.[151] Self to Cotheridge with F.S in aft.

Mon July 9th

Haymaking, hawled 6 loads bank. Took barrel to Ulmnschneider. Borrowed S. Beard's [152] horse and trolley.[153] Fine day.

Tues July 10th

Haymaking in Bank.

Wed July 11th

Finished hawling out of Bank. Mowers cutting orchard. Light heifer from Chambers' cow to Mr Denovan's bull. Took Mr Woodyatt's[154] tedding machine home. [155]

Thurs July 12th

Haymaking orchard and pulling rick at home.[156] Began mowing middle meadow at Common Hill. John fetching 56 gallons beer off Ulmnschneider. To J. Miles paid for half ton hay had for Major Lavie. Sold F.S three heifers and keep for £46.

Friday July 13th

W. Wood[157] fetching load sawdust off Adams and ½ ton hay for Mr Whinfield off J. Miles. John tedding Common Hill and hawling hay in orchard.

Sat July 14th

Mr Whinfield paid £6-15-0 for 1 ½ ton hay and Revd Curtler £2-10-0 for half ton and mare one day. Paid Wilson repair trap £2. Borrowed 30s off Mr Gallagher.[158] Put in O.B £6-15-0. Unloading 2 loads out of orchard onto rick. Discharged men about 11.00 a.m. wet. Paid mowers for orchard and middle meadow Common Hill £2-6-0. Stopt then mowing. To Worcester in aft. Brought back grocery and paid T. Abell 3s for shoeing nag.

Sunday July 15th

Church a.m. To Burrough in aft. 3 heifers and calfs came from Mr Wall's for F. Smith.

Mon July 16th

Haymaking. Finished hawling orchard at home and hawled 2 loads from middle meadow Common Hill. Revd Curtler had nag mare to go to Salwarpe.

Tues July 17th

Haymaking Common Hill. Hawled 1 load, drizzling all day. Fetched 9 cwt coal from Fernhill Heath. Took and erected poles at Common Hill.[159] T. Cook's mare 1 shoe and harness repaired at Adams.

---

151 Looking for the men to hire to mow the grass.
152 Stephen Beard, a market gardener and grocer of 24 Cornmeadow, Claines.
153 A four wheeled horse drawn goods carriage.
154 Charles Woodyatt of Spellis Farm, Hindlip.
155 A horse drawn machine for turning (tedding) hay.
156 Pulling refers to the use of the rick poles and horse fork, where a horse would pull ropes to elevate the fork load of hay onto the rick.
157 William Wood, a general labourer of St Stephen's Terrace, Claines.
158 Henry Gallagher, a Fruiterer of 30 Lowesmoor, Worcester.
159 Rick Poles for making a rick of hay, using a horse fork.

Wednes July 18th

Drizzling. John taking 28 gall barrel to Ulmnschneider. Haymaking Common Hill, hawled 4 loads middle meadow. Revd Curtler had nag to go to Worcester.

Thurs July 19th

½ day haymaking, hawled 3 loads out of middle meadow and 2 out of 9 acre at Common Hill. Wet in morning. To Ombersley after mowers at night. F. Smith paid £4 being difference from keep and 3 yearlings and 2 heifers and calfs purchased of same. Meat bill settled up till July 14th 1883. To Mitre Oak.[160] Order ½ pot potatoes off J. Herbert, 2s.[161]

Fri July 20th

Haymaking about 3 hours Common Hill. Came on wet about half past 10 o'clock. Revd Curtler had nag mare.

Sat July 21st

Wet. John in front garden. To Common Hill in morning. Worcester afternoon. Paid haymakers. Bought grocery and Ann came till Monday. Paid Gallagher £1-10-0 and sent £1 to Stafford.

Sunday July 22nd

Church with Ann a.m. Hindlip in afternoon.

Monday July 23rd

Sold two calfs to F. Smith from heifers bought off same, £4. Haymaking part day. Came on wet about 11-0 am. 28 galls beer off Ulmnschneider and 1 bushel Indian corn off Holthams. John fetched same. To Shrub Hill station in afternoon with Ann and Jessie. Sent them to Wolverhampton.[162]

Tues July 24th [163]

Haymaking ½ day Common Hill. Rose to Hindlip.[164]

Wednes July 25th

Haymaking Common Hill.

Thurs July 26th

Haymaking Common Hill

Fri July 27th

Haymaking Common Hill. Finished hawling, discharged two men.

Sat July 28th

Finished unloading two loads hay Common Hill and topping up rick. Discharged men. To Worcester in aft for grocery and had to hind shoes and two removed to nag mare. Rose home from Hindlip.

xxx Ann[165].

---

160 The Coaching Inn at Crossway Green, Hartlebury.
161 James Herbert, Holy Claines Farm.
162 Most likely to visit Ann's mother at Bilston, Wolverhampton
163 Rose's 12th birthday.
164 There are several references to both Rose and Ann spending time at Hindlip and Isaac rides there often. There were large numbers of staff employed at Hindlip House, and Hindlip Court Farm. Mrs Jowett ran her laundry at Hindlip and Isaac is also likely to have had other friends in Hindlip.
165 Three kisses.

Sun July 29th

To Common Hill in a.m. and Northwick p.m. Mr Cook fetched mare away.

Mon July 30th

John cleaning bits, self to Worcester fair. Paid Jos Miles £2-0-0, for half ton had for Mr Whinfield.

Tues July 31st

Cleaning trap, cutting thistles. To Common Hill in aft and took strawberry Chamber heifer to Mr Denovan's bull for 2nd time.

AUGUST

With the hay made, the hay ricks are finished and thatched. New crop hay is delivered to Isaac's customers and he begins settling up for the beer and for monies borrowed during hay making. The first apples from the Orchard are picked and sold on to Mr Gallagher, the Fruiterer. He settles the last two months bread bills with Mr Wall, the baker. One of his cows gives birth to two dead calves.

Isaac, as a volunteer cavalryman with the Worcestershire Yeomanry, prepares his cavalry equipment and starts attending summer drill at Pitchcroft.

Isaac finds time for some rides out and trips to the local Inns, a night at the Theatre and also a two night excursion to Portsmouth. He is called for Jury duty and attends but finds he has been stood down.

Wed Aug 1st

To Partington to try and buy cow.[166] John taking F. Smith wagon home. W. Price 1 day drawing straw, pulling rick at home. [167]To Worcester station with Rose. Gave Ann towards wages 30s.

Thur Aug 2nd

Pulling and topping hay rick Common Hill. Paid for T. Abell 2 days work at same 6s and discharged him.

Fri Aug 3rd

John fetching load saw dust from Adams, Whatmore thatched rick at home.[168] Self mowing lawn.

Sat Aug 4th

To Worcester after grocery and paid 2s for shoeing nag mare. John mowing thistles, cleaning harness.

Sunday Aug 5th

Common Hill in morning, Northwick at night. Limpey[169] cast two dead calfs 8 weeks before time.

---

166 Frederick Partington of Trotshill Farm, Warndon, formerly of Tolladine Farm, Tolladine Road.
167 William Price, an agricultural labourer
168 All of the ricks were thatched with straw when completed to provide a weatherproof roof.
169 Assumed to be the name of a cow.

Mon Aug 6th

John mowing thistles. Cleaning cavalry tack,[170] cutting stuff for thatching at Common Hill and taking same to Northwick.[171] F. Smith borrowed mare,

Tues Aug 7th

John and self to P Cartridge[172] after 12 cwt straw to thatch rick Common Hill. Self to Worcester in aft for cavalry jacket and to change horse with F. Smith.

Wednes Aug 8th

John at gate post in garden and self to Common Hill after old Black horse, T. Cook, to go to Railway station. Ann and Jess came from Wolverhampton. Drill on Pitchcroft.[173] Paid Wall 2 months bread.

Thurs Aug 9th

John mowing thistles, Common Hill. To Fernhill Heath in aft. Ann took old black horse to Hindlip.

Fri Aug 10th

John fetching 11 cwt coal Fernhill Heath. Whatmore thatching rick Common Hill. Paid him 10s for same and one at home. To Theatre at night. 2 bushel Indian corn off H and Walker at 4/3 bushel. Self washing trap.

Sat Aug 11th

John and self Common Hill. Taking down poles, cleaning up and bringing Ann home. Took Wagon, Mr Barnard's, to Old Hills with Ann, bought back grocery.

Sun Aug 12th

Northwick at night

Mon Aug 13th

To Worcester fair. Paid Mr Perrins[174] £2-7-0 less 10s for keep of dog. 5 cwt hay off Cullis,[175] 2 ½ to Whinfield, 2 ½ to Reverend Curtler. John hawling manure to Pump Meadow. T. Cooke old black horse went.

Tues Aug 14th

John hawling manure to pump meadow. Took Cook's mare to Common Hill, putting up poles. Self ride in evening on nag.

Wednes Aug 15th

John and self trimming hedges. Ann went home, had 10s towards wages. John to Worcester after Cavalry jacket straps and self to Common Hill in afternoon.

---

170 Isaac was a member of the Worcestershire Yeomanry Cavalry (Queens Own Worcestershire Hussars following Queen Victoria's visit to them in 1887). This was a volunteer mounted light cavalry force and had one week's permanent duty each year for which they received a bounty. Drills were held at Pitchcroft (now Worcester racecourse) and Powick Hams.

171 Cutting rick pegs, or spars from willow, to peg the straw thatch in place with. The Willow tree still stands in the bank field today.

172 Phillip Cartridge, a farmer of 156 acres at Upper Smite Farm, Hindlip.

173 Mounted cavalry drill with the Worcestershire Yeomanry Cavalry.

174 Henry Rose Perrins, a veterinary surgeon of Shaw Street, Worcester, also Veterinary Surgeon to the Worcestershire Yeomanry Cavalry.

175 George Cullis, a publican of St Georges Lane, Claines and a hay and coal dealer at Lowesmoor.

Thurs Aug 16th

John and self trimming hedges. Drill on Pitchcroft. W. Wood cutting 18 cwt hay out of rick at home. 2 new shoes put on nag at Adams.

Fri Aug 17th

John taking two in calving cows to Partington and bringing two barrens[176] back and fetching 5 cwt hay from J. Miles for Mr Whinfield.

Sat Aug 18th

John taking 15 trusses of hay from home to Revd Curtler, making total had of half ton, and 15 trusses hay to Mr Whinfield making total had of 15 cwt. Paid Pater 11s borrowed in haymaking. Settled up with F. Smith for calfs had, meat and received off him 17/6 meat paid for up to this day. Paid Banks for rick sheet £1 and paid Mr Day 5/6 for straw had. Paid Wm. Wood for cutting hay 4s. Gave John towards wages £3-0-0 making total now had of £6-12-0. Mr Barnard paid £4-0-0 for milk bill. 2nd horse of S. Beard to tack.

Sun Aug 19th

Black heifer beasting. Church a.m. Common Hill p.m.

Mon Aug 20th

John at hedge. To Worcester fair. Mr Baldwin paid 19s for 5 cwt hay and had 6 trusses more. Paid Mr Adams £1-7-0 and Ulmschneider for beer £5-0-0. Received off same and J. Newey £1-0-0 towards trap. To Raven at night with J. Miles and F. Smith.

Tues Aug 21st

John at hedge

Weds Aug 22nd

John at hedge. Ride with Pater and S. Perk,[177] Cradley and Malvern.

Thurs Aug 23rd

John mowing thistles Common Hill. Bought Cooks mare up from Common Hill. To Worcester in afternoon with spring cart for rubble. Could have none.

Fri Aug 24th

John trimming hedge to Common Hill. Mr Whinfield paid £7-0-0 for keep of horses and 15cwt hay. Drill on Pitchcroft.

Sat Aug 25th

John fetching ½ load saw dust off Adams. Put in OB £9-0-0. Mr Barnard paid £3-0-0, owes 8s. To Worcester, new hat and shirt 9s-0d.

Sun Aug 26th

To Northwick afternoon.

Monday Aug 27th

9 galls beef off Ulmschneider. T.Cook took his mare away. John trimming hedges and taking 2 pots apples to Gallagher for Pater. Went by excursion to Portsmouth. Pater bought Hills wheat £16-17-0.

---

176 Barren cows, unable to produce a calf.
177 Samuel Pearkes, who had farmed Daniels Farm at Tinkers Cross, Claines. A retired single farmer, in 1881 he was boarding with Isaac John's parents, Isaac and Hannah and was also there in 1891. He died in January 1892 aged 86.

Tues Aug 28th

At Portsmouth. John at hedges and taking Chambers' heifer to Denovan's 3rd time. Strange pony got in field.

Wednes Aug 29th

Home from Portsmouth. Dr Hill up to look at mare.[178] John at hedges.

Thurs Aug 30th

John at hedges. Went to Worcester for jury, postponed, not got post card in time.[179]

Fri Aug 31st

John taking mare to have 2 hind shoes put on. Picking ½ pot apples down in bank. Wet afternoon.

---

178 Hilary Hill, MRCS, LAC, surgeon to the Worcestershire Yeomanry Cavalry, surgeon and registrar of births and deaths, 48 Broad Street, Worcester.

179 Called for Jury duty.

SEPTEMBER

Isaac buys in a field of wheat from Ombersley Road and a rick is made from the trusses. Apple picking is now in full flow with red apples and Princess Pippens being picked and sold. Manure is being spread on the grass fields.

The Michaelmas half year rent is due on 29th and Isaac makes several trips to Worcester with his father to try and speak to his Landlord about reducing the rent. He goes for some afternoon and evening rides to Westwood Park and to Malvern with his mother. Having tried to get Mr Chambers' heifer in calf three times with Mr Denovan's bull he takes it to Mr Hillman's bull.

Isaac gets his Cavalry saddle re-lined and attends more drill practice, culminating in the Worcestershire Yeomanry camp, drill and review parade at Powick Hams, as well as undertaking some rifle practice. He receives his annual bounty payment of £2-15-0 from the Yeomanry.

Saturday Sept 1st

John spreading manure in pump meadow. To Worcester shopping and put in O.B £3-9-4. Took ½ pot apples to Gallagher 2s-6d. Wet. 2 bushels Ind. Corn off N.S.W

Sunday Sept 2nd

Wet, to Northwick at night

Monday Sept 3rd

To Worcester fair showed Dr Hill nag mare. John washing trap to ride to Hartlebury at night. Paid J. Herbert 2s for potatoes had.

Tues Sept 4th

Hawling wheat bought off Hill in Ombersley Rd.

Wednes Sept 5th

Finished wheat hawling. S. Beard had 1 pot Prince Pippins & ½ pot red apples.[180]

Thurs Sept 6th

John to Merrimans Hill loading hay for F. Smith. To Drill on Powick Ham.[181]

---

180 A dessert apple. Originally known as Stanardine. In 1832 the young Princess (later Queen) Victoria visited Tenbury and during her stay was presented with a basket of Stanardine. She was reputedly so impressed that a message came back instructing that they be known from then on as the Princess Pippin. (Three Counties Traditional Orchard Project)
181 Drill with the Worcestershire Yeomanry Cavalry.

Fri Sept 7th

John at hedge. To Partington's to see about cows, one has calved. Drive in aft round Westwood Park & Droitwich.

Sat Sept 8th

John at hedge. Self to Worcester in aft. Put in O.B £2-10-0.

Sunday Sept 9th

Church a.m. Northwick p.m.

Mon Sept 10th

To look at cows and calf at Mr Wookey's.[182] John fetching two cows from Partington at tack. Partington bought calf of one for £2-0-0. Shooting rifle butts in aft.[183] Paid Serj. Hill 5s for club money.[184]

Tues Sept 11th

John fetching 10 cwt coal from Fernhill Heath, 7s. Thatched wheat rick sides. Getting up pump in meadow.

Wednes Sept 12th

John in front garden. To Worcester with Pater to see Mr Curtler, no go, then to O.B.

Thurs Sept 13th

Jno[185] picking fruit in corner of bank, about 4 pots. To White's with pump bucket out of meadow. Self to C. Herbert's for to see about half ton hay & to Mr Hill's sale in Northwick Lane.[186] Rev Curtler had nag mare.

Fri Sept 14th

See T.G.C about rent being lowered and iron hurdles. 11 fowls to G. Hughes for £1, paid same time. Took 4 pots apples & ½ pot red same to Gallagher. Rev Curtler had nag mare. Took barrel to Ulmschneider for Pater. Bought back 1 cwt bran off Gallagher & 1 bushel oats, 1 bushel beans from N.S.W for nag mare. New pair shoes off Pott for Jess.

Sat Sept 15th

John cleaning chaff cutter and trimming hedges. To Worcester & paid Holtham £1-3-0 and new tile 7s. Bought two barrels beer from Ulmschneider. Took Cavalry saddle to be lined.

Sun Sept 16th

To J. Miles in a.m. with F.S. Paid him £1 for 5 cwt hay. To Northwick in afternoon.

Monday Sep 17th

To Northwick after pea straw. Topping up wheat rick. John trimming hedges. Cutting thistles Common Hill.

---

182 Richard Wookey, of Warndon farm, farming 231 acres.

183 Most likely to be rifle practice with the Worcestershire Yeomanry Cavalry, at their rifle range at Gorse Hill, off Tolladine Road, Worcester. At that time they were using Snider breech action carbines which turned out to be very unsatisfactory.

184 John Hill, Serjeant Major of Worcestershire Yeomanry Cavalry, Brook Cottage, St Georges Lane South, Claines.

185 An abbreviation for John.

186 Possibly Henry B Hill a farm manager and farmer who retired to Merrimans Hill, Worcester.

Tues Sept 18th

John fetching ½ ton hay from Chas Herbert for Major Lavie £2-2-6. Washing trap. Self took Partington's two cows home. To Malvern with Mater. Whatmore thatched wheat rick.

Wednes Sept 19th

To Worcester fair to see Mr Curtler about lowering rent. John taking 14 trusses hay to Mr Baldwin, makes total had ½ ton. Had two shoes put on nag at Adams. W. Wood cutting hay at home. Red cow cherry calved heifer calf week before time.

Thurs Sept 20th

John & self taking Chambers' heifer to Mr Hillman's bull, being 3 times to Denovans. Drill on Pitchcroft.[187]

Fri Sept 21st

Drill Powick Ham. John at hedges, cleaning out places.

Sat Sept 22nd.

Drill Powick. John taking light heifer bought off F.S to Hillman's bull, no go. To Worcester in aft after grocery. Paid W. Wood towards cutting 2 tons hay at home 5s.

Sun Sept 23rd

College a.m. Northwick p.m.[188]

Mon Sept 24th

Drill on Powick Ham. John to Worcester after Whiskey & Partington had calf from red cow.

Tues Sept 25th

Drill, Powick Ham. John at hedges. 2 trusses hay to Rudge, 4s.

Wednes Sept 26th

Drill Powick. John to Worcester aft, after mare, cleaning tack. Self paid Ostler 5s.[189]

Thurs Sept 27th

To Worcester in morning to try and see Curtler, no go. To Northwick afternoon. John at cows cleaning harness & picking up fruit.

Fri Sept 28th

John taking black and white heifer to Hillman's bull. Getting up potatoes and to Worcester with pot Prince Pippins to Gallagher.

Sat Sept 29th

John getting up potatoes. Self to Worcester after grocery and paid Mr Bill £2-6-6 for flour and received £2-15-0 off Sergeant for Yeomanry.[190]

---

187 *"The regiment assembled for Permanent Duty at Worcester on the 20th September, and was inspected by Colonel Mussenden at Powick"*. (The Yeomanry Cavalry of Worcestershire 1794-1913).

188 It is not known to what "college" refers. It is possible it is linked to Barbourne House (now Gheluvelt Park) which established a boy's college (Barbourne College) in 1883, run by a W.P. Caldwell. (H.W.Gwilliam, Old Worcester). Given all the Yeomanry drill practice it possibly was something to do with that.

189 Ostler= a man employed by an Inn to look after guests horses. Possibly Isaac was stabling his horse rather than going daily to Powick.

190 Payment of the "bounty" for the weeks permanent duty with the Yeomanry.

Sunday Sept 30th

Rudge had 2 trusses hay. Paid Whatmore 3-6 for wheat rick. Northwick at night

OCTOBER

The apple picking continues and the first cider of the season is made. Farm repairs begin again and Isaac is carting soil from the Turnpike in Ombersley Road and sorting out some iron hurdles for the farm. He takes in some pigs to keep for the butcher and receives payment for having let local cricketers use his fields in the summer.

He begins buying in Brewer's grains again and buys a load from the New Inn at Claines who brew their own beer. He purchases a new saddle and bridle; new boots for himself and Jessie and orders a set of new clothes.

The perry pears are ready for picking in the orchard and Isaac, his labourer John and a nameless stranger spend ten days getting them. Unfortunately the fruit picking ladder he has borrowed from his neighbour, Mr Webster of Church Farm, gets broken.

Hedge trimming starts, as well as the regular cutting of hay from the rick and delivering this to Isaac's customers.

Unsuccessful in his attempts to reduce the rent, Isaac pays the second half yearly rent to his Landlord which is now overdue.

Mon Oct 1st

John hawling dirt from Turnpike Road. To Worcester to go look cow and calf with F.S. Did not go, went to see his hay, agreed to take 1 ton for half ton of ours. Put in O/B £5-0-0. Paid baker Wall £1-5-0 two months bread. [191]

Tues Oct 2nd

John finished dirt in Turnpike Road and fetched iron hurdles from T.G Curtler. F. Smith sent up 6 pigs to keep for 1s had per week, him to find flour for same. Took back 3 trusses hay towards half ton. Mater up afternoon.

Wednes Oct 3rd

John and self taking 5 pots fruit to Adams. Made 30 galls cider. Trimming hedges. Preece fetched two Langsham 6s.[192] Albert paid £2-0-0 for use of field for cricketing.

---

191 Charles Wall, baker, 39 The Tything.

192 The Langshan is an ancient breed of hen originating from the Langshan region in Northern China. They were imported into Britain by Major F.T. Croad in 1872.

Thurs Oct 4th

John and self picking up fruit and taking 3 pots apples to Gallagher at 3s pot. To Northwick in afternoon. Bought saddle and bridle from Curtis.[193]

Fri Oct 5th

To Worcester station with Annie. Gallagher paid £1-14-6 for fruit had less 6-6 for 1 cwt bran. John taking 1 ton hay to Dr Hill £3-15-0. Self fetching 14 cwt coal from F. Heath at 12/6 ton. Martin Cole trimming hedge at 1 ½ pearch.[194]

Sat Oct 6th

John fetching 14 cwt coal from F.Heath, cleaning lawn. M. Cole had 4s for trimming hedge. To Worcester for grocery. Rudge had two trusses hay 4s. Paid P. Cartridge for 12 cwt straw, 37s. Paid Insurance money 22s. S. Beard paid 6s for 1 ½ pots apples. Paid John £2-8-0 wages due 29th Sept.

Sunday Oct 7th

To Burroughs with F.S in afternoon.

Mon Oct 8th

John and self getting fruit, large red apples. F.S sent up cow bought in market, Potter's, sold same big white one for £18.

Tues Oct 9th

John and self getting apples, large red ones about 18 pots, broke Mr Webster's ladder. Gave Martin Cole turning bury 4/6.[195]

Weds Oct 10th

John and self getting fruit. Mr Baldwin paid £2-0-0 for half ton hay and Rev Curtler £2-0-0 and £1-5-0 for use of mare. John fetching 9 bushel grains from New Inn.[196] Martin Cole trimming hedge 2/6.

Thurs Oct 11th

John taking nag mare to Adams shoed all round. Boots for repair to Tyler. Getting Prince Pippins. Gave Mrs Barnard ½ pot.

Fri Oct 12th

John to Adams for load saw dust, getting fruit. Paid Carruthers 3s for grains had. Barnard borrowed nag mare.

Sat Oct 13th

John and self getting fruit. To Worcester for grocery and new boots off Tyler 7/6. Put in O.B £11-7-6. Sold part pot Blenheims[197] to Newbury 4/6.[198]

Sunday Oct 14th

Church a.m. Northwick p.m.

---

193 Frederick Curtis a well-known Sadler and harness maker of College Street, Worcester.

194 An agricultural labourer from Pershore Lane, Martin Hussingtree

195 Turning the muck heap (bury).

196 The New Inn, Ombersley Road, spent grains from brewing.

197 The cooking apple, Blenheim Orange, found at Woodstock, Oxfordshire near Blenheim in 1740.

198 John Newbury, a fruiterer of 41 The Tything (West side).

Monday Oct 15th

John fetching 10 hampers from Gallagher and 1 cwt bran from Holtham.[199] Getting fruit picked and started Blenheims. Miss Goldey up in aft. [200] To Worcester at night with her. W. Wood cutting hay 16 cwt out of rick at Common Hill, began same.

Tues Oct 16th

John washing trap, putting down fruit. Self taking 6 pots to S. Beard, paid £1-2-0 for same, same time. To Greaves sale at Ombersley with S. Beard. John picking up spotted red apples in afternoon.

Wednes Oct 17th

John and self getting fruit and taking 11 pots to Gallagher at 3s pot, 33s paid same time. 4 trusses hay to Rudge making total had of 5 cwt, paid same time 19s for it. Light cow Partington calved, heifer calf.

Thurs Oct 18th

John and self getting apple tree in Castle's meadow, turned off about 14 pots. To Worcester put in O.B £4, and ordered clothes off Hansher £3-8-0. Theatre with Mrs G at night.

Fri Oct 19th

John and self taking 5 cwt hay to Rudge & 5 cwt to Baldwin from Common Hill.

Sat Oct 20th

John getting pears, cleaning up front garden and bought mare to Worcester for self. Paid Mr Curtler £133-8-0 half years rent due Sept 29th.[201] Agreed with F.S changed white cow for red one for keep of pigs difference.

Sun Oct 21st

To Northwick at night. John here in aft. Revd Curtler had nag mare.

Monday Oct 22nd

John and self getting red roller pears, sold same to Cupper[202] 3s per pot.[203] Mr Barnard paid £5 milk. Sold calf from Partington for £2-5-0, sold 4 trusses hay to man for 8s from Common Hill. Miss Gold up. Revd Curtler had nag mare.

Tues Oct 23rd

John and self taking white cow and 3 calves to Common Hill. Picking roller pears. Revd Curtler had nag mare.

Wednes Oct 24th

John and self picking pears and stranger ½ day at same. Mr Cupper had 10 pots of red roller pears at 3s. Revd Curtler had nag mare. To Webster's sale. Paid C. Herbert £2-0-0 for half ton hay had off same Yeomanry week.

---

199 Hampers = wicker baskets measuring a pot.

200 Miss Ann Elizabeth Goulde, a dressmaker of 9 Droitwich Road, Claines.

201 The second half yearly farm rents were traditionally due on Michaelmas Day (St Michael's day 29th September).

202 John Cupper, who worked for Frederick Barnard the Dairyman.

203 A small round pear for perry (pear cider). Possibly a local name for the ancient Herefordshire Red Pear variety also known as Blunt Red, Red Horse, Sack.

Thurs Oct 25th

Stranger and John 1 day picking red roller pears. 8 bushel grains New Inn, 4/6. Paid same time. To Worcester races. Revd Curtler had nag mare.

Fri Oct 26th

Stranger ½ day picking pears and John at same. Revd Curtler had nag mare. To Worcester races.

Sat Oct 27th

John and stranger getting fruit. Paid stranger 6s for 3 days work. Mr Barnard paid £4-0-0 towards milk money. Put in O.B £8-18-0. To Worcester after grocery and paid Ulmschneider for 3 barrels beer, £1-10-0. Paid grocer 2 weeks 13s-3d. Rev Curtler had nag mare. Rode F.S old doctor home.

Sunday Oct 28th

Revd Curtler had nag mare. Church a.m.

Monday Oct 29th

John taking light Wall heifer & Smith heifer to Taylor's bull.[204] To Worcester after Annie. Bought Jessie boots 4/6, to Old Hills at night. Took half pot Princes Pippins to Mrs Jeavons, 3s-0 paid same time. W. Wood cutting 18 trusses hay Common Hill paid same, 2s. Revd Curtler had nag mare.

Tues Oct 30th

John taking half ton hay to Revd Curtler from Common Hill. Picking up apples. Mr Cupper fetched 20 pots red roller pears at 3s pot, paid and 3 pots Prince Pippins at 6s pot. Took 4 pots to Ombersley for same. Revd Curtler sent nag back. To Common Hill to see calves.

Wednes Oct 31st

John getting pears. To Northwick a.m. Revd Curtler had nag mare at night to go to Worcester. 1 peck apples to Warner's man for 1s.

---

204 Daniel Taylor of Danes Green Farm, Claines.

NOVEMBER

The final pears are picked and the cows are allowed back into the Orchard to graze. The cider machine arrives and around 120 gallons of cider are made. 50 gallons of this is sold to Reverend Curtler for six shillings. Isaac makes enquiries about taking on a new labourer.

Isaac rides out with the Worcestershire Hunt at Hindlip, gets himself a new suit of clothes and attends his Theatre night.

The winter routine of buying in bran and feed continues and Isaac takes Oats to be ground at Hawford Mill. Hay and Apple sales are made and these are delivered around the parish. Mr Barnard settles his hay bill which has been outstanding for four years. Isaac makes several payments into his bank.

Coal is bought in for the winter, being collected from the Railway station at Fernhill Heath. Isaac takes on a new boy labourer, John Powell. Mr Chambers' heifer is still not in calf, having been served four times so Isaac takes it to Mr Taylor's bull.

Thurs Nov 1st

John getting pears. Revd Curtler sent nag back. To Miss Jowett's with Pater. Theatre night, gave Ann £1 towards wages.

Fri Nov 2nd

John finished getting red roller pears. 3 trusses hay from home to Turner & 2 trusses (Common Hill) to F. Smith, total now had by F.S 7. F. Smith sent up cow and calf for T.G.C to look at. Cows turned 1st time in Orchard.[205] Coachman Revd borrowed mare at night.

Sat Nov 3rd

John cleaning up front garden. To see T.G.C about cow and calf, not right sort. To Worcester after grocery. Put in O.B £5, see Mrs Firkins[206] about new boy. John brought mare down night. New suit clothes off Hansher £3-8-0.

Sun Nov 4th

To Common Hill & Northwick. F. Smith had 1 truss Common Hill hay, total had towards ½ ton, 8 trusses.

---

205 Turned in after picking the fruit.
206 Possibly Elizabeth Firkins, storekeeper, 14 Northwick Road, Claines

Monday Nov 5th

John taking mare to Adams, shoed all round. W. Wood cutting hay, Common Hill. Hunting Hindlip. Mr Barnard paid £5 owes £1 same.

Tues Nov 6th

John at cows. Wet day. Self taking half ton (Common Hill) hay to Mr Whinfield. 1 cwt Common Hill hay to Turner, 4s. Paid same time. Colley brought cider machine and began making.[207] Gave Raisen 2s. Fetched 2 barrels from father, Northwick.

Wed Nov 7th

John, self, Colley making cider, 3 ½ hogsheads.[208] Paid Colley 17s for same. Sold 50 gallons to Revd Curtler at 6s and gave 50 gallons to Pater. 2 trusses Common Hill hay to F. Smith, making total had 10 trusses towards half ton. W. Wood cutting hay, Common Hill. Took cider to Northwick.

Thurs Nov 8th

John cleaning up yard and raspberry bed in garden. To Common Hill. W. Wood cutting 2 ½ tons hay. Paid him 10s for same. Left down in shed just 2 ton.

Fri Nov 9th [209]

John and self taking 1 ton hay to Mr Whinfield & 5 cwt hay to Mr Baldwin & 10 cwt hay to Major Lavie from Common Hill, & 1 truss hay to F. Smith total had 11. Self brought back 1 cwt bran from Holthams & 1 peck oats off Hughes 1s, paid same time. Mater up in afternoon. Received cheque off Dr Hill for £3-15-0 for 1 ton hay had from home in October.

Satur Nov 10th

John taking roan cow Partington to Mr Taylor's bull. To Common Hill, 1 truss Common Hill hay to F. Smith. To Worcester, took 1 pot Princes Pippins to Gallagher. Paid Hansher. Sent £3 to Stafford. Paid Henry Griffiths tar sheet. Put in O.B £5.

Sun Nov 11th

At cows. Northwick in aft. 1 truss hay (Common Hill) to F. Smith.

Mon Nov 12th

John fetching load rubble from gas works.[210] Brought half bred Alderney cow off F. Smith for £15-0-0. 1 truss hay to F. Smith, from Common Hill. To Worcester fair.

Tues Nov 13th

John fetch 2 sacks oats from Worcester, T. Cook. Mr Barnard paid £3 for hay had 4 years ago. 1 truss Common Hill hay to F. Smith. Took black and white cow to Taylor's bull, turned from Hillmans.[211] Mr Warner ground 4 bushel oats.[212]

Weds Nov 14th

John & self taking calf from half Alderney bought off F. Smith to Mr Jackson for £2-4-0. To Worcester in afternoon. John went to Miss Jowett's. 1 truss Common Hill hay to F. Smith.

---

207 Most likely a scratter for mashing the apples and a press for the juice.
208 Hogshead barrel containing 54 gallons.
209 Isaac's 42nd birthday.
210 In Tolladine Road, Worcester.
211 Turned = would not mate with.
212 William Warner, a miller of Hawford Mill, Claines, now Hawford Mill campsite. See photographs.

Thurs Nov 15th

John and self taking 1 pot Princes Pippins and part pot Russets[213] to Mr Aaron. 10 cwt Common Hill hay to Quarterman's £2 paid same time. 10 cwt Common Hill hay to T.G.Curtler £2 and 1 truss Common Hill hay to F. Smith. W. Wood cutting hay at Common Hill. 1 cwt potatoes off Coroner.

Fri Nov 16th

John and self repairing fence Common Hill. 2 trusses Common Hill hay to F. Smith. W. Wood cutting hay Common Hill

Sat Nov 17th

W. Wood cutting hay Common Hill. 3 trusses hay to F. Smith. John cleaning front garden. Self to Worcester, paid T. Cook £1-1-0 for 8 bushel oats & M. Walker £1-8-0, & paid W. Wood 10s for cutting hay.

Sun Nov 18th

John at cows. To Common Hill & Northwick, F. Smith had 3 trusses hay.

Mon Nov 19th

John to Common Hill & to Bourne's after withys for pot sticks.[214] Mr Cupper fetched 22 pots Princes & 18 pots large red apples at 6s per pot, £12. Mr Barnard paid £5-10-0. 1 truss Common Hill hay to F. Smith. Paid W. Wood 2/6 for hay cutting, total paid him 12/6 for 3t, 4 cwt cut at Common Hill. Ride to Fernhill Heath on mare.

Tues Nov 20th

To Worcester, took 10 trusses Common Hill hay to F. Smith and gave his cattle 1 truss at Common Hill, making total had this day 11, & took 4 trusses Common Hill hay to Turner & 10 cwt Common Hill hay to Mr Whinfield. Paid Fuller 6/9 for pork.[215] Took tin for repairs & bought new kettle 1/3. Aaron paid 10/6 for apples. Put in O.B £16-10-0.

Wed Nov 21st

John fetching load rubble, at cows and cutting thorns. Paid Wall baker Sept bread 12/3. Gave Ann 5s towards wages, going to Hindlip and Worcester. Coroner sent 16 cwt straw to T.G Curtler Esq.

Thurs Nov 22nd

John to Worcester after 40 bushels Vinegar grains, Coroner had 20 of them. Paid Smith ½ cwt potatoes 3/6. Bought two Dandy brushes, new file 6/2 repaired milk tin 1/6.[216] Mr Turner paid 8s for 4 trusses of Common Hill hay. Revd Curtler had nag mare to Worcester. 1 truss hay to F. Smith.

Fri Nov 23rd

John Common Hill and at cows, gapping in Captain Castle's Meadow. To Hindlip with apples for Miss Jowett, and to Worcester for grocery. T.G Curtler paid for half ton of hay and half ton straw had off Coroner £4-4-0. Paid Powell, Cannock Chase £1 for coal. 1 truss hay to F. Smith.

Sat Nov 24th

John at cows and washing yard and bought nag down at night to Worcester. Paid F.S for half bred Hereford cow and calf 17/10. Mr Jackson paid £2-4-0 for calf. Put in O.B £6-4-0.

---

213 One of the russet apples, but unlikely to be the now popular Egremont Russet which was not recorded until 1872.
214 Sticks of willow, from which the wicker pot measures were made. Bourne, at Camp Meadow Farm was next to the River Severn so presumably well supplied with willow.
215 George Fuller, labourer at Hawford Farm, adjoining Oak Farm.
216 Dandy = Brush for grooming horses.

Sunday Nov 25th

Common Hill. Chambers' heifer to bull, Mr Taylor, been 3 times to Denovan's and once to Hillman's before.

Monday Nov 26th

John trimming hedge, Pump Meadow and Orchard. To Worcester on nag for ride. T. Cook sent up old horse for chaff cutting, no use.[217] Bought slippers off Tyler 2/6.

Tues Nov 27th

John fetching in half ton hay from Common Hill to sell at home. To Common Hill on nag in afternoon, bought 1 Truss Common Hill hay to Tupper.[218]

Wed Nov 28th

F. Barnard clipping nag mare. New boy John Powell came. Took Strawberry cow to Taylor's bull second time.

Thurs Nov 29th

John at cows & John Powell and self to Worcester after grains from Vinegar works. 20 bushels & Woodyatt had 20 of same 8s/4. Bought grocery 7s/3. Sold Langshan pullets to Herbert Turner 7d. Revd Curtler had nag mare. F. Smith sent up fresh cow for her milk.

Fri Nov 30th

Powell and John in front garden. I took 3 trusses Common Hill hay to Turner, paid same time. Revd Curtler had nag mare.

---

217 Implies the presence of a horse gin, horse driven gear for driving barn machinery.
218 Lewis Tupper, a Railway Clerk, living at Bevere.

DECEMBER

Isaac dismisses his former labourer, John Shuck, pays him his wages due but deducts the cost of the fruit picking ladder from his wages, which was broken in October. Winter days are interspersed with rides out to Worcester and Hindlip and hunting at the Old Hills and at Cutnall Green.

The threshing machine arrives at Oak Farm and the rick of wheat which he bought in is threshed out with the help of seven men, including his 72 year old father. As Christmas approaches Isaac settles his newspaper bills, sells cider and makes some preparations for Christmas, including lending money to Mr Price and his father. He gets his father half a gallon of whisky and gives his mother a sow.

He finishes writing the diary for 1883, two days before the first anniversary of his late wife's death.

Sat Dec 1st

Powell at cows. I paid John Shuck two months wages less 4/6 for ladder broke and discharged him.[219] To Worcester in afternoon had nag mare shoed all round at T. Able's, paid same time. 2 trusses of Common Hill hay to F. Smith out of 23 left in shed.

Sun Dec 2nd

1 truss Common Hill hay to F. Smith.

Mon Dec 3rd

Gave Ann going to Worcester 2s. Jno Powell[220] hawling hedge brushings at cows and took three bushels of Oats to be ribbled to Warner.[221] Ride on mare in afternoon to Worcester. 2 trusses of Common Hill hay to F. Smith.

Tues Dec 4th 1883.

Jno Powell and self fetching 1 ton 6 cwt coal from Fernhill Heath at 12/2 ton. Ride to Hindlip afternoon. 2 trusses of Common Hill hay to F. Smith.

Wednes Dec 5th

Jno hauling manure from Revd to Orchard. 1 truss Common Hill hay to F. Smith.

---

219 See October 9th.
220 Jno=abbreviation for John.
221 Ribbled oats were oats which were slightly crushed with an indented roller. Driven by hand or horse.

Thurs Dec 6th

Jno hauling manure from Revd Curtler. 30 bushels of grain from Vinegar works, J. Beard had 10, paid same time. To Worcester and Northwick at night. T.G Curtler had old horse. T.Cook went up for chaff cutting, £5 to be price.

Fri Dec 7th

Jno at cows washing trap and to Worcester after grocery, milk tin repair, new saucepan, bass broom and T.G Curtler sent £5 for T. Cook's horse. Brought Mrs Jeavon's barrel.

Sat Dec 8th

Jno at cows, washing yard and W. Wood hay cutting in Common Hill, 6 trusses left in shed out of 23, F. Smith had 17 of same. W. Wood cut 44 and put in shed today. To Worcester in afternoon. Barnard paid £2 towards milk.

Sunday Dec 9th

To Northwick a.m.

Monday Dec 10th

To Worcester, borrowed Ned Hughes' mare.[222] Took 30 gallons cider to F. Smith and 9 gallons same to Mrs Jeavons who paid 6s for same. Gallagher 6/6 for pot princes. Mrs Turner paid for 2 trusses hay took today 4s. Paid Mr Denovan 18s for cows to bull.[223] Lent W. Price 1s making 2 now lent. Lent F 2s.[224] Paid W. Wood 4s for cutting 44 trusses hay. Fetched back back-strap off harness from Curtis.[225]

Tues Dec 11th

Jno at cows, I was hunting Old Hills. Threshing machine put down.[226] Gave Ann 4s towards wages.

Wednes Dec 12th

John Powell cutting chaff, self at cows. 1 day spreading manure and digging in garden. Bought shovel off Tommy Hughes for 6d.

Thurs Dec 13th

Jon fetched 40 bushel grains from Vinegar works. I took two trusses Common Hill hay to Lavendar, not paid for. Paid grocer 12/6. Self at cows.

Fri Dec 14th

Jno, Self, T. Hughes, J. Cox, J. Doughty, Jack threshing, F wheat rick.[227] 20 sacks, paid Barnes[228] for same £1-5-0. And 4 men 2/5 each.

Sat Dec 15th

Tommy ½ day cutting chaff and cleaning up front garden. John and self at cows. Barnard paid £1 towards milk. ½ gallon of Whiskey to Pater. 2 trusses hay from Common Hill £1.

---

222 Edwin Hughes, a farmer of 20 acres, Egg Lane, Claines.
223 Payment for bull hire.
224 Father.
225 Leather back strap, part of the harness that takes the weight of the shafts.
226 Arrival of the threshing machine, to thresh the wheat from the straw stored in the rick since harvest. It is not clear whether this was steam powered, or via a horse gang and gin wheel.
227 Father, presumably loading from the wheat rick.
228 David Barnes, a Brickfields farmer and subsequent "Threshing Machine Proprietor"

Sun Dec 16th

Church a.m. Northwick p.m.

Mon Dec 17th

John and self at cows. To Bills Mill and Worcester for ride after. Paid Herald[229] office 2/6, Tommy ½ day.

Tues Dec 18th

John and self taking 5 cwt Common Hill hay to G. Thomas Pack £1 and bought load sawdust off Adams and taking 17 cwt 1 gr straw to Quarterman at 50s ton. 1 cwt bran off Bill. Tommy ¾ day loading straw.

Wed Dec 19th

John at Cows. Self hunting Cutnall Green.

Thurs Dec 20th

John and self fetched 40 bushels grains from Vinegar works. Self 2 trusses hay to Lavendar and 1 half ton to Quarterman Common Hill Hay. Gave mater 1 sow. T.Cook sent up horse. T. Hughes ½ day. Had out of Common Hill shed 30 trusses of 67. F.S had sent half Alderney bulling.

The diary ends

21. The last known photograph of Isaac by the Granary at Oak Farm, c: 1928

229 The Worcestershire Herald newspaper.

Oak Farm Accounts 1883

Income[230]	**£**	**s**	**d**
Hay	49	15	9
Milk	65	1	9
Horse Hire	12	13	0
Livestock sales	278	7	11
Fruit	23	4	6
Cider	0	17	$1^{1/2}$
Keep	8	17	6
Miscellaneous	15	13	3
Yeomanry	2	15	0
Total Income	**£457**	**5s**	**$9^{1/2}$d**

Farm Expenses	**£**	**s**	**d**
Livestock purchases	107	14	0
Livestock feed	48	0	10
Bull Hire	1	0	6
Seed	0	4	0
Threshing	1	5	0
Repairs	8	4	2
Blacksmith	1	19	6
Farm Wages	24	7	2
Rent	265	8	0
Beer & Cider	6	19	6
Miscellaneous	5	10	6
Total Farm Expenses	**£470**	**13s**	**2d**
FARM PROFIT/LOSS	**-£13**	**7s**	**$4^{1/2}$d**
Household Expenses			
Meat	11	6	$7^{1/2}$
Groceries	7	1	$2^{1/2}$
Bread	4	17	3
Coal	4	8	6
Clothing & Boots	12	16	5
Childcare & Schooling	9	5	6
Grave expenses	7	6	0
Papers	0	2	6
Wages to Ann	4	12	9
Total Household Expenses	**£61**	**16s**	**9d**
Grand Total Expenses	**£532**	**9s**	**11d**
PROFIT/LOSS AFTER HOUSEHOLD EXPENSES	**-£75**	**4s**	**$1^{1/2}$d**

230 £1 from 1883 was equivalent to £111 in 2016 (Bank of England inflation calculator).

POSTCRIPT

What became of the main people mentioned in the diary?

Isaac John Sansome (self): (1841-1930). Isaac John married his wife's cousin Ann, who had been housekeeping for him, in April 1884. They went on to have five children: Annie, John Isaac (Jack) twins May & Lily and Harry, my grandfather. Isaac continued to live at Oak Farm, with his second wife, and after her death, Harry and his family. He died there in 1930, aged 89 and was buried in Claines churchyard in the same grave as both of his wives.

Ann: Annie Greenshill (1856-1913). Annie was the cousin of Sarah Ann, Isaac's first wife. She was housekeeping for Isaac and the diary more than hints at a developing romance with Isaac John in 1883. They were married April 2nd 1884 at Claines Church by Revd. Alfred Porter, witnessed by her cousin Fanny Cope and her brother Charles Greensill. She bore five children and died at Oak Farm, 20th January 1913 aged 57.

Rose: Rosamond Elizabeth Sansome (1871-1903). The first child of Isaac John and Sarah. Rose was 11 years old when her mother died. Rose was never spoken of by any of the family, either by her true sister Jessie, or half-brother Harry. It was only when researching the family tree in the 1980's that she was discovered. It transpires Rose gave birth to an illegitimate daughter in 1888 when she was 17. The daughter died at the age of five months old and is buried in Astwood Cemetery, Worcester. Rose emigrated to the United States and settled in Toledo, Ohio. She was married twice, first to Ernest Macon, was widowed and then to Albert P Van-Allen. Rose died in December 1903, aged 32 years, the same age as her mother, Sarah Ann.

Jessie: Jessie Hannah Sansome (1880-1955). The second child of Isaac John and Sarah and the first Sansome to be born at Oak Farm. Jessie trained as a nurse and worked at Powick Asylum, then in Bristol and Norwich. She married Tom Palmer at Claines in 1917. They settled in Cowbridge, Glamorgan, where she died with no children.

Pater: Isaac Sansome (1812-1898). Isaac John's father. Born in Uplyme, Devon. He married Hannah Pippen in Axmouth in 1837 and they moved to Worcester in 1839. He worked as an agricultural labourer for Richard Spooner, Esq, MP at Brickfields Farm, then farmed at Rushwick Farm, Worcester; Moat Farm, Astwood (The Perdiswell Estate) and retired to 24 Northwick Terrace (12 Vine Street in 1891), Northwick. He died at Oak Farm in 1898.

Mater: Hannah Sansome (1810-1899.) Isaac John's mother. Born Hannah Pippen in Axmouth, Devon. Died at Oak Farm in 1899.

What became of Oak Farm?

Isaac John continued farming at Oak Farm. In 1890 the 30 acres of land at Common Hill were given up to a Mr Arthur James Beauchamp. In exchange Isaac John took on an 8 acre field known as Claines Meadow. His youngest son Harry, who was wounded but survived the First World War, took on the tenancy of the farm 24th September 1918, buying his father's share of stock and machinery for £314-19s. In 1923 Harry married Leah Lenora Packman, the granddaughter of the Fishmonger and Fruiterer William Packman, and they had five children at Oak Farm. These were Enid, Peter, Stanley, Harry, Maurice and Leah. Harry senior continued farming at Oak Farm until his death in 1982, all that time a tenant of the Curtler family. At that point his eldest son, Peter, purchased the farm from the Curtler family as the Bevere Estate was finally broken up. Peter lived at Oak Farm until he died in 2013, with the express wish that the farm continued in the family, as it does today.

22. Descendants of Isaac John Sansome, at Oak Farm,
on the occasion of his Grandson Peter Sansome's 80th Birthday, 2005

23. Replanting the Orchard at Oak Farm in 2017.
Left, Harry Packman Sansome 88; right Stanley John Sansome 89, Isaac's grandsons.
Centre, Great Grandson, Geoff Sansome

ANNEX 1. List of names mentioned in the Diary

Name in diary	Details
Aaron, Mr	Henry Aaron, Inn keeper of the Crown & Anchor in Hylton Road, Worcester
Abell, T	Thomas Abel, a Blacksmith of Bransford Road
Adams	Charles Adams, a Blacksmith of Ombersley Road
Adams	George Adams, Landlord of the Lansdowne Inn, Claines
Albert	Unknown, possibly of the White Hart Inn, Fernhill Heath
Alford, J	Unknown
Amphlett	Edmund Amphlett, a druggist of Mealcheapen Street, Worcester
Baldwin	Henry Baldwin, a hay customer and market gardener of Ombersley Road
Ballinger, Mr	Unknown
Banks	Henry Banks, a rope manufacturer, 36 The Shambles, Worcester
Barnard	Frederick William Barnard, Dairyman, Bevere
Barnes	David Barnes, a dairyman and farmer at Brickfields who in 1901 became a "Threshing Machine Proprietor" in Station Road, Fernhill Heath.
Bayliss	Unknown
Beard, S	Stephen Beard, market gardener and grocer of Cornmeadow, Claines
Bill, Miller	Henry Bill, miller at Mildenham Mill, Claines
Bourne	Thomas Bourne, farmer at Camp Meadow Farm, Bevere
Bowden	John Bowden, carpenter of Ombersley Road, Claines
Bridges	Unknown
Burroughs	Farmer, unknown, possibly of Broadheath
Burrows	Possibly Edward Burrow, who farmed at Spring Bank, Claines (1861) and then retired to Hallow
Butler	Unknown, general builder and labourer
Cartridge, P	Philip Cartridge, farmer at Hindlip
Caruthers	Thomas Carruthers, Landlord of the New Inn, Claines
Castles, Captain	Captain Charles Castle, of Hawford House, Deputy Lieutenant of Worcestershire
Chambers	Farmers at Astwood Road, Claines
Coachman, Revd	Joseph Smith of Bevere, the coachman to Revd. Curtler
Cole, Martin	Martin Cole, an agricultural labourer of Pershore Lane, Martin Hussingtree
Colley	Henry Colley, a labourer of Lowertown, Claines
Cook/Cooke, T	Thomas Cook, Horse dealer, 39 Bransford Road, Worcester
Cosfords	The Drapers, 57-60 High Street, lived at London Road, Worcester
Cox, J	Unknown, a labourer helping with the threshing
Crisp, G	George Crisp, of Cosfords Tailors, London Road
Cross, Aunt	Sarah Cross, of Stretton, Cannock. An Aunt to Isaac's first wife Sarah, through the Cope family
Cullis	George Cullis, publican, St Georges Tavern, St Georges Lane, Claines and a coal and hay merchant of Lowesmoor Wharf, Worcester
Cupey	Unknown
Cupper	John Cupper, servant to Frederick Barnard, the dairyman
Curtis	Frederick Curtis, Sadler, College Street, Worcester
Curtler, Revd	Reverend Thomas Gale Curter of Bevere Knoll, Vicar of St Stephen's Church and son of Thomas Gale Curtler, Isaac's landlord

Name in diary	Details
Curtler, Thomas Gale (TGC)	Thomas Gale Curtler (Snr), of Bevere House, owner of Oak Farm and the Bevere Estate
Day	William Day, a farmer at Tolladine
Dean, Mr	George Dean, the gardener and Bailiff of the Perdiswell Estate, Claines
Denovan	James Denovan, a farmer, Holy Claines Farm, Claines
Doughty, J	Unknown, a labourer helping with the threshing
Evans, Mr	Possibly Charles Evans & Co, iron and brass founders, St Martin's Gate Worcester
Firkins, Mrs	Possibly Elizabeth Firkins, wife of Joseph Firkins, storekeeper 14 Northwick Road
Fraser	Unknown, of Ombersley
Fuller	George Fuller, labourer at Hawford Farm, Claines
Gallagher, Mr	Henry Gallagher, fruit merchant, egg importer, herring curer and licensed dealer in game, 30 Lowesmoor, Worcester
Golde/Gold, Miss	Ann Elizabeth Goulde, dressmaker of Droitwich Road, Claines
Greaves, J	Unknown, farmers at Ombersley
Gresley	Reginald Archibald Douglas Gresley, JP. Farmer at High Park, Droitwich
Griffiths, Henry	Rope and marquee manufacturer, The Tything, Claines
Hale	William Hale, Grocer, 11 St Swithin Street, Worcester
Halls	J&F Hall, Ironmongers and merchants. 1&2 The Shambles, Worcester
Hancher/Hansher	Michael Hamsher, tailor, draper and military outfitter, 3 St Nicholas Street, Worcester
Harris	William Harris, Market Gardener, Cornmeadow, Claines
Herbert, C	Charles Herbert, a farmer at Tapenhall Farm, Claines
Herbert, J	James Herbert, a farmer at Holy Claines Farm
Hill	Unknown. A Farmer/Market Gardener in Ombersley Road, Claines
Hill, Dr	Hilary Hill, MRCS, LAC, surgeon to the Worcestershire Yeomanry Cavalry, surgeon and registrar of births and deaths, 48 Broad Street, Worcester
Hill, Serj.	John Hill, Serjeant Major of the Worcestershire Yeomanry Cavalry, Brook Cottage, St Georges Lane South, Claines
Hillman	Thomas Hillman, a farmer at Hill House, Great Witley
Holland	Alfred Holland, a tailor of St Georges Lane, Claines
Holthams	The Seed and Corn merchants in the Cornmarket, Worcester
Horton	John Horton, a farmer at Suddington, Ombersley
Hughes (Coroner)	William S.P Hughes, County Coroner and farmer at Northwick Hall
Hughes, Edwin (Ned)	A farmer at Egg Lane, Claines
Hughes, Tommy	Thomas Hughes, a labourer from Dilmore Lane, Fernhill Heath
Jeavons	Thomas Jeavons, a plumber & painter and family friend, Cumberland Street, Claines
Jowett, Miss	Jane Jowett of the Laundry, Martin Hussingtree
Larkworthy	Agricultural Engineers at Lowesmoor, Worcester
Lavendar	Unknown, possibly connected to Jane Lavender & Mrs Gutch of Barbourne who gave the land for the building of St Stephen's Church
Lavie, Major	Major Ernest Lavie, JP of Bevere Cottage, Claines
Marshall	George Marshall, a labourer of Ombersley Road, Claines
Miles, Jos	Joseph Miles, a farmer of Moat Farm, Astwood, Claines
Morton	James Morton of Hadley, Ombersley
Mrs G	Unknown
Newbery	John Newbury, a fruiterer at 41 The Tything, Claines

Name in diary	Details
Newey, J	James Newey, a market gardener and farmer at Severnside, Northwick, Claines
Old Bill	Unknown, a labourer
Pack, G Thomas	Unknown
Packman	William Packman a fishmonger and fruiterer at 11 The Cornmarket, Worcester
Partington	Frederick Partington, a farmer at Trotshill Farm, Warndon
Payne	John Payne, a farmer at Chatley, Ombersley
Perk, S	Samuel Pearkes, a retired farmer who lodged with Isaac John's parents.
Perrins, Mr	Henry Rose Perrins, a Veterinary Surgeon of The Butts, Worcester, Veterinary Surgeon to the Worcestershire Yeomanry Cavalry. Lived at 45 Foregate Street, Worcester
Powell, Cannock Chase	Coal Merchants, Cannock, Staffordshire
Powell, John	A farm labourer at Oak Farm from November 1883
Preece	Unknown, a customer
Presdee, Old	William Presdee, of Crosspools Cottage, Claines
Price, W	Possibly William Price, an agricultural labourer, born at Claines 1841 and now living in Ombersley
Quarterman	John Quarterman, a butcher of Droitwich Road, Claines
Raisin	John Raisen, the gardener to the Whinfield family at Severn Grange, Claines
Robinson	Possibly Joseph Robinson, farm bailiff to RP Hill, Esq of Brickfields Farm, Claines
Rudge	Henry Rudge, a milkman of Crown Lane, Claines
Shuck, John	The farm labourer at Oak Farm until December, discharged and wages deducted for breaking a ladder!
Smallwood, Revd	William John Smallwood, Vicar of Claines
Smith, F (F.S)	Frederick Smith, a butcher of 43, The Tything, Claines. His father had farmed 300 acres at Hindlip Court Farm, as well as being a butcher. His mother runs the butchers shop in the Tything after her husband's early death in 1854
Smith, J	Unknown, a farmer at Hallow, Worcester
Stearman, Miss	Fanny Rose Stearman, a Governess of a Ladies boarding and day school, Ivy House, Sansome Walk, Worcester
Stock, Mrs	Unknown, a labourer
Taylor, Daniel	A farmer at Danes Green farm, Claines
Tearne, J	Jack Tearne, a labourer
Thornbury	Jesse Thornbury, a stone mason of Northwick Place, Claines
Tomley	Unknown, a labourer
Tupper	Lewis Tupper, a railway clerk of Bevere, Claines
Turner, Herbert	Unknown, a customer
Tustin	John Tustin, a farmer and horse breaker, Ronkswood.
Tyler	William Henry Tyler, Bootmaker, St Martin's Gate, Worcester
Ulmnschneider	Charls Ulmnschneider, Landlord of the Five Ways Inn, Angel Place, Worcester
Walker, M	Unknown
Wall	William Wall, a farmer at Ladywood, Martin Hussingtree
Wall, Baker	Charles Wall, a baker in the Tything, Worcester
Wall, T	Unknown
Warner	William Warner, Miller at Hawford Mill, Claines
Webster	Anthony Webster, a farmer at Church House, Claines

Name in diary	Details
Whatmore	William Whatmore, a thatcher of Tinkers Cross Cottages, Claines
Whinfield, Mr	Edward Wrey Whinfield, of Severn Grange, Claines
White, HT	Henry White, a blacksmith in Droitwich Road, Fernhill Heath
Wilkes	R&H Wilkes, builders, St Georges Lane North, Claines
Wilson	John P Wilson, a farmer at Firs Farm, Bevere
Wood, W	William Wood, a labourer of St Stephen's Terrace, Claines
Woodyatt, Mr	Charles Woodyatt, a farmer at Spellis Farm, Hindlip
Wookey, Mr	Richard Wookey, a farmer at Warndon Farm

ANNEX 2. Index of photographs with credits

All watercolour illustrations of farm animals. John Coleman, Claines, 1998.

Front cover, Isaac John Sansome, 1900 and the bank at Oak Farm, 2017.

Back cover, the bank at Oak Farm, 2017. G. Sansome

THE FARMING COMMUNITY NETWORK

Farming has and always will be a risky occupation with farmers having to contend with many variables outside their control such as the weather, animal disease and fluctuations in the market prices for the inputs they require and the crops and animals which they grow. Add to this the normal pressures of family and rural life and their deep passion for the land on which they work and farming families can face a complex tapestry of issues which can, at times, seem almost impossible to overcome.

In Isaac John Sansome's day, farmers were very independent and tenacious, choosing to keep their worries to themselves, often doing a great deal of damage to their physical and mental health as a result. These days, farmers are just as independent and tenacious and still have to deal with many of the same issues as did Isaac but with the added complexities of modern farming life: ever stricter legislation governing their day to day activities; seemingly endless layers of bureaucracy to deal with; and a general public which has become detached from the realities of food production yet wants to have an increasing influence over farming practices.

Just as well then that in 1995, in the face of a sustained decline in farm incomes and an alarming increase in suicides amongst farmers, that The Farming Community Network (then known as Farm Crisis Network) was formed. Now a network of over 400 volunteers based in groups all across England and Wales, FCN is a well-respected national charity providing practical and pastoral care to farming families who are suffering stress and anxiety as a result of issues affecting either their farm business or their domestic life. In addition to a telephone helpline, which operates from 7am until 11pm every day of the year, FCN operates and email helpline with a view to arranging for a suitably qualified volunteer to visit and "walk with" the farmer and his or her family until their issues are resolved. This totally confidential and free at the point of delivery service supports around 6,000 individuals every year.

It costs around £1,500 per day to resource our volunteers and run the organisation, so we are very grateful to you for the indirect support you are giving FCN by buying this fascinating book which is a window into a by-gone farming age whose traditions and issues are still remarkably familiar today.

Helpline: **03000 111 999** **e-Helpline:** **chris@fcn.org.uk**

Website: **www.fcn.org.uk**